AMBRA |V

# CULTURAL KNOWLEDGE WORK

# The Politics of Knowledge Work in the Post-Industrial Culture

## Understanding the Dissemination of Knowledge of the Sciences, Humanities, and the Arts

René Stettler

Neue Galerie Luzern and the Swiss Biennial on Science, Technics + Aesthetics
University of Applied Sciences and Arts, Lucerne, Switzerland

AMBRA |V

Author
René Stettler
Neue Galerie Luzern http://www.neugalu.ch
The Swiss Biennial on Science, Technics + Aesthetics



© 2014 AMBRA | V
AMBRA | V is part of Medecco Holding GmbH, Vienna
Printed in Germany

Layout and cover design
Livia Gnos, Montreux, Switzerland

Cartoon on cover
Gabi Kopp, Lucerne, Switzerland

Copy editor
David Matley, Switzerland

Printing and binding
Strauss GmbH, Mörlenbach, Germany

Printed on acid-free and chlorine-free bleached paper

With 49 colour images, 6 black and white images, 8 diagrams

ISBN 978-3-99043-546-5 AMBRA | V

# Contents

# Foreword

René Stettler has successfully worked for over two decades to establish ongoing publically accessible, dialogue between leading thinkers in the world of arts, sciences and technological development, which contributes to an evolving framework of ethical and social responsibility. His knowledge of the field is deepened by his personal relationship with many artists and scientists of world stature, whose theses and values he weaves into the fabric of his own advanced thinking. His intellectual journey leads us into a future of a fruitful diversity of ideas, attitudes and behaviours, which is both grounded and forward-looking.

The book examines, within a holistic frame of transdisciplinarity, work practices in the knowledge fields, and their effects in the post-industrial culture on public understanding and identification with the arts and sciences. It seeks to provide a utopian vision of an alternative future that might support a contemporary cultural epistemology. René Stettler sees the possibility of cultural change arising from informed public debate, and from adaptations within the ecology of information that might yield new ideas, practices and attitudes. This is a call, both informed and impassioned, for social action, referencing a number of successful projects undertaken by the author.

Within the framework of construction, production and distribution, the book promotes a *second-order* perspective on cultural work that challenges contemporary forms of political power and social control. It is a call for action at the interstices of cultural disciplines, where uncertainty and contingency can be celebrated in the search for new forms of practice, sustainable environment awareness, democratic debate, and enriched qualities of individual and social interaction.

It is a work of profound reflection and intellectual courage, based on thorough scholarship, exemplary ethics and visionary sensibility.

Roy Ascott, August 2013

# Acknowledgements

I owe a great debt to many individuals who helped to make this book possible. First, I am grateful to Roy Ascott, who provided enormous amounts of encouragement and friendly support. I would like to thank Roy for his guidance and advice in the context of my earlier doctoral thesis on the subject of this book. I thank in particular David McMullan and David Turnbull, who both responded unfailingly to requests for assistance. David Turnbull's book *Masons, Tricksters and Cartographers* opened the door to an understanding of our differing ways of producing knowledge and the ways in which knowledge practices work today.

The explorative atmosphere of my sessions with the Planetary Collegium helped me to acquire important insights into and understandings of research in cultural work practices and their role in the humanities and the arts as producers of valuable knowledge and defenders of civilisation. I am especially grateful for conversations with Martha Blassnig, Margarete Jahrmann and David McConville during and beyond the sessions of the Planetary Collegium. I particularly thank Michael Punt and Mike Phillips, who, with their knowledge, experience and creative thinking, have helped to shape my vision. I am indebted to Christina Ljungberg, who gave me valuable insights into other research fields. I thank Christina for her great help and generous support that I have received during discussions on the topic of my research. I want to offer very special thanks to David Matley for the critical and meticulous reading of the entire text.

I am deeply grateful to a large number of people for conversations about science, philosophy and ecology. They include, in alphabetical order, Simon Berther, Reinhold Bertlmann, Dick Bierman, Bob Bishop,

Rainer Blatt, Charlotte and Josef Brandenberg, Fritjof Capra, Ulrich Claessen, Lüder Deecke, Ruth Durrer, Hans-Peter Dürr, Lilly Fellmann, David Finkelstein, Hanspeter Fischer, Ernst von Glasersfeld, Christine and Walter Graf, Giselher Guttmann, Stuart Hameroff, John Horgan, Brian Josephson, Kevin W. Kelley and Rachel Bagby, Ursula and Herbert Kneubühl, Christian Thomas Kohl, Gerhard Johann Lischka, Pier Luigi Luisi, Luis Eduardo Luna, Josef Mitterer, Sir Roger Penrose, Jack Pettigrew, Robert Poole, Karl Pribram, Peter Schulz, Benny Shanon, Abner Shimony, Uli Sigg, Henry Stapp, Andrea Steimer, Axel Vogelsang, Franz Vollenweider, Peter Weibel, Margaret Wertheim, and Judith Wyrsch. My gratitude to these people, who I either invited to the Swiss Biennial on Science, Technics and Aesthetics as speakers or who supported the Biennial in various ways, cannot be adequately expressed. At many of the Biennials, I have had the opportunity to exchange ideas about the nature of life, civilisation, consciousness, social reality, technology and culture, and other issues with these people. I am grateful for these dialogues and stimulating conversations and the great individuals, colleagues and friends that I have met during the past 25 years. They have supported and encouraged me to pursue and develop new concepts and ideas for new conferences, meetings and projects. Many of these ideas are articulated in this book.

I am especially indebted to Otto E. Rössler for many inspiring conversations and feedback of various kinds. I am also indebted to Irena Banjkovec, who supported me in the rather difficult early stages of my research. I am indebted to Markus Allemann, Andy Athanassoglou, Muriel Bonnardin Wethmar, Marcel Brenninkmeijer, Roland Meyer, Kumi Naidoo, Alejandro Roquero, Kaspar Schuler, Verena Vanoli and Verena Weiss for numerous enriching discussions about issues of sustainability, ecology and green politics. I owe thanks above all to Ana Iribas Rudín for many conversations in the spirit of simplicity and heartfelt wisdom.

I am especially grateful to Guy André Mayor and Beny von Moos, who both passed away a few years ago, for challenging philosophical discussions about science, art, culture, mind and life. They both recurrently encouraged me to explore the epistemological and political dimensions of cultural projects, practices, institutions and people as producers and mediators of knowledge in order to make sense of a deeper social and moral-ecological understanding of the post-industrial culture.

And finally, I want to thank my son Pablo, my sisters Gaby and Karin and their husbands René and Beat, and my parents Heidi and Karl, who have both passed away. My family has supported my efforts in many helpful ways and I thank my son, my sisters and my parents for entrusting me with their knowledge, experiences and insights into life domains with which I am not so familiar.

# Note on Terminology

The key theme of this book is an exploration of the potential for a renaissance of *cultural work* and *knowledge* in the global cultural economy. Throughout, the emphasis is on a particular kind of creative cultural labour that links knowledge to the different practices of cultural work which I will colloquially call "cultural knowledge work". While I will define the cultural work and cultural workers I am concerned with in the Introduction, I will set "cultural work" and "cultural knowledge work" in quotation marks in various parts of my text. When "cultural work" is placed between quotation marks, I have reservations about the prevailing practices of cultural work and the function and meaning of cultural work as defined in prevailing theories. "Cultural work" can then be associated with a more open and progressive concept of cultural work practice. "Cultural knowledge work" is mostly set in quotation marks in the first part of my text, but in the later chapters I more often use the italicised version of *cultural knowledge work* without quotation marks. *Cultural knowledge work* (CKW) is a propositional concept that by definition is investigated in this book with regard to the epistemic features of cultural work, and the "epistemology"/ecology of the cultural labour process.

Another focus of this research is concerned with the reconceptualisation of what I recurrently refer to as (techno-)socio-cultural spaces of knowledge. When *techno* is bracketed, *(techno-)socio-cultural* refers to the kinds of cultural, epistemological and political spaces that we construct in the process of targeting audiences with collective matters-of-concern which are framed by the complex socio-economic-technological (post-industrial) reality that we inhabit. When *techno* is *not* bracketed, *techno-socio-cultural* refers more precisely to the systematisation and dissemination of knowledge in spaces/places in which the cultural performance and reproduction of technoscience are particular issues (such as the Gotthard Base Tunnel exhibition; see *Building a Transalpine Railway Tunnel and Manufacturing a Public and Transient Cultural Space of Knowledge*).

*The humanities and arts* are associated with a form of knowledge production in contemporary society. Their scope is broad, concerned with diverse social, cultural and analytical practices which accomplish inherently reflexive and often contemplative tasks—a complex of primarily critical or speculative ideas, methods/disciplines, values, intellectual and social demands as distinguished from the mainly empirical approaches of the natural and social sciences as well as science and technology.[1] In this context, the *humanities- and arts-based dialogues* imply the diffusion of knowledge of this kind as well as the provision of material-semiotic responses to socio-technological and political post-industrial reality in cultural environments.

---

[1] My understanding of *the humanities and arts* overlaps with Michael Gibbons et al.'s definition of the humanities in his work *The New Production of knowledge* (1994), which I have used for this book. For further references on the work of Gibbons et al. see «The Scope of this Book».

# The Scope of this Book

> We are always attempting to retie the Gordian Knot[2] by criss-crossing, as often as we have to, the divide that separates exact knowledge and the exercise of power—let us say nature and culture. (Latour 1993:3)

I was 13 years old in 1968 when Stewart Brand published the *Whole Earth Catalog*. At that time, I was too young to understand that this was—alongside the landing on the moon a year later and the paramount aesthetic revelation of seeing the Earth, the "Blue Marble", from a distance—the beginning of the "environment". 40 years later, this whole environment—our home planet—is at stake due to climate change and environmental destruction, which is potentially disastrous for life and civilisation. Our linear pursuits and activities of economic growth and material consumption, as Fritjof Capra argues (2002:208), are causally connected to global warming, climate change and the changing life conditions on the planet. These issues are now scientifically evidenced and no longer hypothetical.[3] Thus, my highest ambition is that this book should contribute to a theoretical framework of legitimate considerations in which reflections of how to modify our anthropology of creation may add to a finer sense of possibility for the cultural production and dissemination of knowledge, the redefinition of the *politics* and *ecology* of cultural work, and the exploration of a practice-based "epistemology".

---

[2] *The Gordian Knot* is a legend associated with Alexander the Great. It is often used as a metaphor for an intractable problem solved by a bold stroke, "cutting the Gordian knot".
[3] Scientists who study global warming and climate change are currently in a state of suppressed panic. Things seem to be moving much faster than their models predicted. In *Klimakriege* (2008), the German sociologist Harald Welzer criticises the blindness of the social and cultural sciences with regard to the expected dramatic consequences of global warming.

I grew up in a small village at the foot of the Alps near the city of Lucerne in the heart of Switzerland. During my adolescence, my emotional and intellectual awareness of the whole gamut of the fragile conditions of life and human culture was impacted and formed by the unreconciled political atmosphere of the Cold War and the threats of a nuclear war between America and the former Soviet Union. Having thus experienced my childhood and adolescence, in a sense, as an ontological uncertainty and a threat to the societal and cultural collective in which I was raised, I had as a boy a recurring dream of my village being invaded by foreign military forces. Moreover, the Catholic environment with its rich cultural tradition and symbolic figurative world made me experience myself as embedded in processes between materiality and semiosis, a very deep involvement in the reality of symbolic and sacramental spheres, religious doctrines of incarnation and trans-substantiation.[4] As a consequence, my understanding of and relationship with the concepts of "human", "nature", and especially with "human nature", "human culture" and the universe were deeply entangled in an awareness of a small and subtle matrix of "high stake balancing acts" (Haraway 1997b:47) in a morphing environment— a struggle between the physical and the non-physical, the organic and the non-organic, the material and the semiotic, the epistemological and the political.

In 2008 when I started working on my earlier thesis on which this book is based, I was diagnosed as having Lyme disease. Lyme disease or *borreliosis* is an emerging infectious disease caused by a species of bacteria (*spirochetes*). The diagnosis coincided with the beginning of the crisis of the world's financial systems, impacting the world's economy and

---

[4] I owe the awakening of memories and forgotten experiences in my Catholic childhood to Donna Haraway. On several occasions, Haraway recalls the experience of Catholicism at a young age which had a deep impact on her intellectual development (Haraway, Goodeve 1997b:54-55; Haraway 1997a).

markets on a global scale. The sudden realities of abused fiscal responsibility by banks, global economic recession, unemployment etc. revealed the timeliness of investigating in greater depth the significance of the topic of my book, that is, to explore the role of the humanities and arts as producers of new knowledge and defenders of civilisation. The biological conditions and global economic and political frameworks in which I was embedded—the complex networks of nature and the convoluted networks of human society—led me to reflect on the kind of taken-for-granted order of the contingent rhizomes to which my existence and that of other beings belonged. I began to make sense of Capra's (2002) under standing of the connections between life, mind and society—the hidden web of life, consciousness and social reality. Furthermore, the thinking of R. Buckminster Fuller[5]—one of the most important American thinkers of the 20th century and visionary for the 21st century—his fascinating mix of utopian vision and organic pragmatism, his credo of "more for less" and his belief in the interconnectedness of all things encouraged me to pursue many of my thoughts that are articulated and developed in this book. It was R. Buckminster Fuller's work which sensitised me to the issue of closing the gap between the natural sciences, technology, the humanities and the arts.

My encounter with *borreliosis* framed the writing process and my entire intellectual journey. Going through the process of healing, that is, eliminating or rather inhibiting the radius of power of the *spirochetes* in my body, gave me valuable insights and food for thought. My interest in the 3.5-billion-year-old bacterial pathogen leading an asymmetrical war against my life tempted me to keep track of biological and physical metaphors as I started to take tremendous pleasure in structural-functional complexes

[5] Richard Buckminster Fuller (1895–1983) was an American architect, designer and visionary.

at microscopic levels.[6] During the medical treatment,[7] I developed a respectful attitude for the inscrutability of natural phenomena, and the complexity of the interplay between human activity and ecological processes.


## A New Communal Space for Knowledge

A central aim of this book is to contribute to the concept of a new communal space for knowledge—a social and symbolic space with a whole new ecology in which diverse interests, knowledges and values can co-exist (cf. Turnbull 2000). The spectrum of my interests and themes thus began to evolve around the following key questions:

- To what type of future public cultural "spaces" of knowledge and "models" of traffic for knowledge should we aspire?
- What future cultural work practices are needed to sustain these "spaces", and the ontological, epistemological, moral and ethical dimensions that our being-in-the-world entails?
- How should the contingency of human experience be addressed in domains of the cultural production and dissemination of knowledge?
- What answers to these questions support the retying of the Gordian Knot?

These questions—I am tempted to say—are rooted in my childhood and adolescence. They have "materialised" or rather ripened in my mind

---

[6] As a "true super-microorganism" borrelia bacteria outwit the body's immune system and change its function by using deceptive strategies and camouflage; they survive temperatures of up to minus 50 degrees Celsius (Storl 2007:51, 52). Spirochetes' strategy is to "excrete antibiotic toxins" or to "adopt cystic form while they fall into a long sleep until the biological environment has improved" (Ibid.:27, 51)

[7] I took *doxycycline*, Pfizer's first once-a-day broad-spectrum antibiotic, but then followed Storl's recommendations by using a plant named *Fuller's Teasel (Dipsacus fullonum)* for which a number of medical properties, though not proven in medical trials, are claimed, among them curing Lyme disease, antibiotic properties and improved circulation.

in the past 50 years, and today express my commitments and very personal concerns with the contemporary problems of rationality, reflexivity, transdisciplinarity, globalisation and ecological thinking, including those relating to the rational and authoritative forms of knowledge that my people and my culture have created.

The work of a number of authors in science, philosophy, sociology, the history of ideas and other disciplines have contributed thoughts and themes to this book. The Czech-born media critic and philosopher Vilém Flusser has been a key figure with his seminal view of the future of design, cultural ethics and our ways of "designing" the world. Flusser's (1990) criticism of the "stubbornness" and the "non-creativity" of politics, which, in his view, is based on "outdated categories" such as "nuclear weapons", "energy crisis", "distribution of goods" or "Third World", has accompanied me throughout the writing process. A number of authors have encouraged me to pursue ideas across the borders of academic disciplines. The work of these authors with a heart-and-mind concern with ethical issues spurred me to combine different approaches, and to address the contemporary challenges of cultural, social and environmental renewal. The late Heinz von Foerster, who introduced epistemological doubts to cybernetics, thus confounding the mechanistic ideas held by early cyberneticists, impressed me with his legendary enthusiasm and unforgettable vitality during public lectures. I was fortunate to witness one of his public talks on the occasion of the 1992 Berne conference *Der entfesselte Blick,* organised by Gerhard Johann Lischka.[8] Von Foerster's ethical and aesthetic imperatives, which focus on seeing ourselves as a part of the universe, as *participants* rather than outside observers, have influenced my thinking as much as Fritjof Capra's insights into social and economic pitfalls in *The Hidden Connec-*

---

[8] Gerhard Johann Lischka is an Austrian-Swiss media theorist and editor.

*tions* (2002). Capra's concepts of how to build ecologically sustainable communities provided many insights and helped me to understand how to implement sustainability in civil society through cultural work. The work of Flusser, von Foerster, Capra and especially that of the Austrian non-dualist philosopher Josef Mitterer made me aware of the tendentious dichotomies inherent in our political and epistemological representations.

However, the main sources that have inspired me in writing this book, and that I have used in order to make sense of a deeper under-standing of contemporary culture and knowledge, are, first, the writings of the French historian Michel Foucault and those of the French sociolo-gist Bruno Latour. A brief introduction to the work of Latour and Actor-Network Theory, including its significance for the theoretical framework of this book, will be presented in the Introduction. Many of the ideas of these authors concerning the forces and human behaviours that drive and determine science, politics and the economy have contributed to my text. The work of Gregory Bateson in *Mind and Nature* (1979), and Humberto Maturana's and Francisco Varela's guide to the formation of cognition and human intelligence in *The Tree of Knowledge* (1987) have added an incentive to understand the biological roots of the human constitution, and our capacity for ethical reflection, deliberation and decision-making. Michael Gibbons et al.'s work *The New Production of Knowledge* (1994), which explores the changes in the ways in which scientific, social and cultural knowledge is produced, supported me in some of my ideas concerned with the future of the cultural production and dissemination of knowledge. It also encouraged me to rethink esta-blished cultural practices and policies. The European Commission (EC) report *Taking European Knowledge Society Seriously* (2007)[9] on key

---

[9] The producers of the report are Ulrike Felt and Brian Wynne. The highly-qualified international expert group consisted of Michel Callon, Maria Eduarda Gonçalves, Sheila Jasanoff, Maria Jepsen, Pierre-Benoît Joly, Zdenek Konopasek, Stefan May, Claudia Neubauer, Arie Rip, Karen Siune, Andy

issues concerning science and governance in the EU has sensitised me to diverse policy concerns and to new ways of addressing the problems we face today with "reflexive" thinking. The report encouraged me to explore my concept of "cultural knowledge work" (CKW) and supported me in attempts to define a more sustainable cultural life. By proposing an ethics of freedom, Michael Hardt's and Antonio Negri's work *Commonwealth* (2009) has given me insights into a possible constitution for our common wealth. It supported me in focusing on the problem of how to articulate the vision of a global commonwealth adequately from the perspective of the cultural workplace. *Commonwealth* has also challenged and enriched my thinking concerned with the biopolitical production of knowledge in the age of globalisation. Ulrich Beck's examination of the 21st century risk society, drawing together the new world order where terrorism, financial turmoil and global climate change haunt our lives and engender new risks, has given me valuable insights into the human situation and a cosmopolitical perspective from which to reflect on our self-inflicted problems. Finally, a valuable resource has been Mark Banks'[10] *The Politics of Cultural Work* (2007), in which he provides an overview of the intellectual traditions that appear to be supporting emergent empirical studies of cultural work. Cultural work has been substantially neglected in the literature as a topic of research. In his work Banks offers a broad theoretical understanding of cultural work from the perspective of "critical theory" approaches ("Frankfurt School" Marxism)[11] and a "governmental" ("neo-Foucauldian") approach that sees cultural work as a mode of managerial "business-oriented" authority.

Stirling and Mariachiara Tallacchini. In order to make the collective authorship of the 14 personalities more transparent, I will refer to it in this book as the work of Felt, Wynne et al.

[10] Mark Banks is a British researcher and author with interests in cultural work, media culture, cultural policy and urban space.

[11] Critical theory approaches such as the Frankfurt School have investigated cultural work from the scope of social conditions such as the problematic alienation of cultural workers from the possibility of "authentic self-formation", as Banks points out (2007:182), and thus stress the impact and the consequences of the cultural industries, exploitation of workforce and economic globalisation etc.

## Grasping the Socio-Technological Post-Industrial Reality

Overall, my text is concerned with the transient, epistemological, and political dimensions of cultural work, institutions and people as authorities and mediators in the business of grasping our socio-technological post-industrial reality. In using the word "post-industrial", I draw on the theories of Daniel Bell (1973) and the taxonomy of George Ritzer (2007). Although the theory of the information revolution may provide a clearer theoretical and empirical framework than the commonly used term "post-industrial society" (examples of post-industrial societies include the US, Canada, Japan and Western Europe), my understanding of "post-industrial" is based on the idea of a society in which an economic transition has occurred from a manufacturing-based economy to a service-based economy, on the diffusion of national and global capital and on mass privatisation. Among several salient changes in social structure associated with the transition to a "post-industrial society", Ritzer's taxonomy emphasises that *theoretical knowledge* as the basic source of innovation is increasingly important in a post-industrial society *instead of* practical and empirical know-how. Advances in knowledge lead to the need for other innovations such as ways of dealing with *ethical* and *moral* questions. Therefore, the exponential growth of *theoretical* and *codified knowledge* is central to the emergence of the post-industrial society. Further, in post-industrial societies new intellectual technologies such as cybernetics, game theory and information theory are developed and socially implemented.[12]

[12] See the conclusions in chapter IV «Epistemology of Post-Industrial Cultural Work» with regard to the discussion of second-order cybernetics as a paradigm for the ethical, aesthetic, pragmatic and political convergences of cultural work.

## Theory of Cultural Work

My final aim is to develop a new theory of cultural work. The book thus ventures into the intentions and motivations behind three selected cultural projects. These projects strongly evoke the extraordinarily heterogeneous complexity of post-industrial knowledge. The first is a public exhibition on the new 57km-long Gotthard Base Tunnel, the longest tunnel in the world, which will be under construction in the Swiss Alps until 2016.[13] The second is Jennifer Baichwal's[14] documentary film *Manufactured Landscapes* (2006)—a portrayal of the Canadian photographer Edward Burtynsky and his work on China's industrial revolution; and the third is Ai Weiwei's[15] social performance *Fairytale* at Documenta 12, for which the artist invited 1,001 Chinese people to the city of Kassel in 2007. My inquiries into the nature, theory and practices of cultural work will address challenges, contradictions, interests and imaginations around these three projects.

[13] The Gotthard Base Tunnel is a new railway tunnel under construction in Switzerland. With a projected length of 57km (35 miles) and a total of 153.5km (95 miles) of tunnels, shafts and passages it will be the longest tunnel (of all railway and road tunnels) in the world upon completion, ahead of the current record holder, the Seikan Tunnel (connecting the Japanese islands of Honshū and Hokkaidō). The tunnel is part of the Swiss AlpTransit project, also known as New Railway Link through the Alps (NRLA), which also includes the Lötschberg Base Tunnel between the cantons of Berne and Valais (Vetsch 2002:152-56; *The New Gotthard Rail Link*. AlpTransit Gotthard Ltd. [Ed.] 2005:2-5, 8-9, 14).

[14] Jennifer Baichwal is a Canadian filmmaker.

[15] Ai Weiwei is a leading Chinese artist.

# I – Introduction

> In the current crisis, there are no backroom villains, no hidden cabal of evil plotters; we are all enmeshed in systems of commerce and manufacturing that perpetuate our problems. (Goleman 2009:39)

For the past 30 years, as a cultural worker and "cultural intermediary"[16] I have practised the social distribution of knowledge by establishing dialogue opportunities and creating platforms for democratic deliberation and debate about science, technological innovation and art. My concerns with the disturbing realities of material consumption, environmental destruction and the present moral crisis are rooted in an awareness that has recently become keenly apparent to many observers. More and more people are beginning to see the structural nature of our problems and of built systems that are based on limitless growth. In *Ecological Intelligence* (2009), Daniel Goleman points to the severe environmental consequences of human activity for the planet, calling it an "onslaught against the natural world" (39).[17] Drawing on the sociologist Angela McRobbie, Banks points to the key problem of market-driven individualisation with its "destructive, desocializing tenor", which promises only further demoralisation and dislocation (2007:163; McRobbie 2002).[18] For John Armstrong[19] the different facets of the crisis are only the "outward signs of a deeper

---

[16] Here I use the term "cultural intermediary" in Bourdieu's (1984) sense.

[17] According to the Earth Policy Institute, there is abundant scientific evidence that humanity is living unsustainably, and returning human use of natural resources to within sustainable limits requires a major collective effort. Sustainability has become a wide-ranging term that can be applied to almost every facet of life on Earth, from a local to a global scale and over various time periods. <http://www.earth-policy.org> downloaded 29 March 2010.

[18] Gibbons et al.'s "market of life-chances" (1994:91) seems to have yielded to the high pressure of the culture industry, whose "nefarious" product, according to Banks (2007:5), is what "Frankfurt School" critical theory has defined as the outcome of an alienating and de-socialising process.

[19] Armstrong is a British writer and philosopher.

malaise" (2009, no pagination). While we should be deeply worried about these signs, we should at the same time start to reflect on new ways of forging a stronger relationship between the humanities and civil society.

## Different Humanities

In this book, I will follow Armstrong's vision of a *different* humanities, one in which the humanities and the arts are the "custodians of a discourse about private and public virtue" (Ibid.). If we assume that cultural work might be on the verge of a new period of intellectual and educational accomplishment, a number of issues such as human excellence and a new ecological-cosmopolitical understanding (Sloterdijk 2010:37) of what we do and how we act pose fundamental challenges for the post-industrial cultural workplace. In order to create an awareness of the consequences of our behaviour, as well as to contribute to more sustainable foundations for a global society and culture,[20] the particular challenge lies in developing a new *compass* that could help us to shift values, "reprogram" our destructive pursuits, and rethink accumulated habits and collective aspirations, as Reto Ringger suggests (2009:20).[21] However, the underlying question regarding the future of social, economic and ethical practices is whether the imperatives of globalised consumer culture and the forces of commodification and marketisation—to which the production and dissemination of

---

[20] I do not attempt to provide much elaboration on the term *culture*, which, especially in recent years, has undergone yet more transformations of meaning. I use the term in a broader sense to refer to ways of life, the arts, the media, political, educational and religious culture, and attitudes to globalisation. For the purpose of this book, I suggest that culture, as it is expressed in the arts, literature, film, and different practices of mediation and representation, draws from and participates in the construction of culture as a system of human values, beliefs and behaviours.

[21] Ringger is a Swiss pioneer of the concept of sustainability investments.

knowledge are vulnerable—have rendered the world incontestable.[22] Different authors have problematised these issues from social, economic, political and ecological perspectives (for example, Altvater 2007; Capra 2002; Castells 2000; Gibbons et al. 1994; Horkheimer and Adorno 1947). While some have suggested "remoralising" social, economic and educational practices in the "search for place" (Banks 2007:145), others have called for future paths towards a renewed ethic of civic virtue.

[22] Banks notes: "Yet, even now, ongoing transformations in the structure of capitalism appear to provide cultural industries with the opportunity to obtain even greater shares of wealth, power, and control (...) [T]he social and economic impacts of globalization now appear to have exacerbated the scale and scope of the cultural 'standardization, commercialization, and rigidification' that Adorno (...) first identified in the high-modern period" (2007:25). In an essay published after his death, Adorno in fact reiterated that the "entire practice of the culture industry transfers the profit motive naked onto cultural forms" (1981; 1991:99), thus anticipating the commercial control of all kinds of social practices that we can observe today.

# The Fragmentation of the Humanities and Arts

Since World War II research and education have focused overwhelmingly on science and technology, which has resulted in the marginalisation of the humanities with negative repercussions for both knowledge and the human community. While the humanities appear to play an ambivalent role in modern culture, their intellectual values are at the same time formed by the social context in which they evolve and are practised, and, according to Gibbons et al., thus become entangled in markets in a "more diffuse sense" (1994:91). Formally and academically, the *humanities* represent our usual human activities as well as the meaning and values that we construct in different ways by looking at the world and thinking about life. Yet, the *work* of the humanities and the *knowledge* which they involve are not one particular thing. Their outcomes and the many forms of knowledge that they provide are all around us and present in our daily lives. They are the newspaper we read, the news we watch on television, or the exhibition we visit in the local art museum. Whether we watch an episode of *Star Trek,* read Stefan Zweig's biography of *Marie Antoinette,* listen to the *Ritual Orchestra and Chants of the Khampagar Buddhist Monastery,* or visit San Francisco's *Exploratorium,* these activities are all part of the humanities. The humanities are always a manner of commitment that embraces human knowledge with a central concern for human beings and our self-conscious awareness of how we know, what we know and how we act. In short, the humanities structure and sustain our views and ways of thinking about the world, our ontological demands and our cultural behaviours.

The new debate about the humanities and arts touches on key issues of how they could forge a much stronger relationship with society in view of their marginal standing compared to the central role of science and technology, and confronts us with difficult questions for which there are no immediate answers. Armstrong's intellectual and ethical stance, which attempts to understand both the causes and conditions underlying the issue, helps to make the core problems in which the humanities (and the arts) are deeply entangled more transparent. Armstrong suggests that

Fig. 1.1

current shortfalls in funding are not the main reason for the growing marginalisation, but instead tend to *obscure* it. He highlights some of the crucial aspects of the relationship between the humanities and the research paradigm in which they are embedded, and notes:

> The core problem can be expressed in the language of economics: the humanities are geared to supply, particularly to supply research. They ignore demand. They are set up to produce refereed journal articles; this is the overarching goal. The impact of research is measured largely in terms of citations in learned journals. This is a fair measure of the regard in which an article is held. However, the audience for these

> journals consists overwhelmingly of other researchers: that
> is, of fellow suppliers of scholarly material. So, what this cri-
> terion of success cannot capture—in fact, what it completely
> disregards—is whether anyone else is interested in, attracted
> to, inspired by, moved by or enlightened by this work. (2009,
> no pagination)

While Armstrong's conclusions address the problem of institutional bias in the humanities towards what he calls the "supply-side conception of the situation", he claims that research needs to be redefined to embrace a "wider range of cognitive virtues" as well as new qualities of thinking such as grace, charisma, intimacy, spontaneity, wit, depth, simplicity, grandeur, warmth, openness, drama, intensity and generosity (Ibid.). For Armstrong, these qualities are linked to both the power of ideas and the ways in which these ideas "get inside our lives and come to matter in everyday existence" (Ibid.). As a consequence, for Armstrong the highest achievement of the humanities is a combination of "scholarly merits" and the wider vision of "intellectual merit" (Ibid.), as we are "stuck with the word research", he claims (Ibid.). Similarly, and in acknowledging the uncertain status of the humanities and arts in the post-industrial culture, Brian Opie emphasises the dominating position of the discourses of science and technology over those of the humanities and arts. He concludes that a similar publicly recognised *term* to promote what the humanities and arts do is missing. According to Opie:

> Major obstacles to shifting the center of gravity of this dis-
> cussion are the lack of appropriate terminology and the frag-
> mentation of the arts and humanities in the modern period
> into media forms with their own institutions, disciplines
> with their separate identities in academic organisational

structures, and government agencies whose status reflects the marginal standing in conventional economic thought and power-knowledge relations of the arts and humanities. By contrast, the term "science and technology" has assumed a generality of reference which allows the institutional and disciplinary differences within science to be subsumed, especially for public discussion, by the one all-encompassing term. A similar term with a similar degree of public recognition is needed for the domain of knowledge represented by the arts and humanities; (...) (2001:2)

The argument for which I seek to provide analysis and evidence in this book is that the engagements of the humanities and the liberal arts are seriously challenged by the continuous growth of science and technology, and therefore poised for a new period of intellectual and educational accomplishment. In the past, discussions of and publications on the curricula of the humanities and liberal arts have raised public awareness of academic criticism of the exclusivity of their traditional canons and methodologies since the late 1960s, yet the political popularity of instrumentalist views of education and employment in particular has at the same time increased, as Berger et al. note (2001:21). The 2000 *Report of the Humanities, Science, and Technology Working Group* published by the American National Endowment for the Humanities, for example, problematises several aspects of the traditionally practised work of the humanities. These include public unease with the social and cultural changes registered by scholars and literary authors in the 19th century that accompanied the growth of science and technology, as well as the growing imbalance between the humanities and the sciences following World War II in the 20th century (Ibid.:3). Over a century later, the issue of public unease with science has remained, but is now expressed in the growing concern with and reactions to issues of environment and health, as Gibbons et al. note (1994:7).[23] The more rapid and large-scale deterioration of the environment, increasing poverty, alienation and social disintegration, terrorism, nuclear proliferation, and global warming coupled with the danger of future climate wars have become daily public issues as many people wish to exercise more control over and have more impact on the outcomes of the processes of research in science and technology. The open question is to what extent the humanities (and arts) will be able to anticipate the consequences of science and technology for human society, and whether the "remoralisation" of social, scientific, economic and educational practices will be successful or not.

The near collapse of the world's financial systems has triggered a new kind of debate about the humanities and arts that is gradually spreading into the public domain. While for many the conduct of the humanities and the liberal arts is on the right track, there are those who argue that our intellectual and moral life is "out of kilter" and that our faith in modernisation, economics, politics and science "no longer rings quite true", as Bruno Latour notes (1993:9, 5). From the perspective of cultural work,[24] which I will explore in more detail in the following chapters, the more challenging question is how to rethink the *kinds* of knowledge that people need and to reflect on the *conditions* under which it ought to be produced and disseminated.

---

[23] Gibbons et al. emphasise the growing public interest in and concern about science and technological risks (notably in relation to nuclear power) as well as the potential dangers associated with biotechnology and genetic engineering (1994:36).

[24] The broad scope of cultural work that I am concerned with is that defined by Banks as the "composing, designing, imagining, interpreting, and manipulating of symbols in order to create music, television programmes, films, art, clothing, graphic designs, images, and other forms of texts" (2007:189). The term "cultural workers" thus includes those experts and scholars who work as cultural "intermediaries" under the mandate of science and governance as well as in other domains such as philosophy, sociology, policy analysis and law, including representatives of the public cultural interest and art or media institutions. As "curators", "arbitrators" and "mediators" in the business of defining and grasping today's social and cultural realities, or as "cultural activists" and "cultural

# The Ambiguity of Knowledge Work

My eating the apple of post-industrial knowledge from within the sociology of knowledge and its constructivist foundations is challenged by the current cultural, economic and political conditions. In addition, the "knowledge society" and the global reach of corporate knowledge economies are a further serious challenge in themselves. While the post-industrial "knowledge society" proposed by Daniel Bell (1973; 1987) has become a topic of research in the sociology of knowledge studies and "knowledge" is now considered a "central resource" that contributes to the productivity of the economy, to the "rationality of politics" and to human orientation in everyday life (cf. Mayntz et al. 2008), there is broad agreement that the role of "knowledge" in any society is that of being the "capital". Economic success therefore involves valorising and harnessing this knowledge. As a consequence, "smart" business organisations are not only foundational to the existence of the knowledge society, which "in its core uses 'knowledge work' to transform the industrial society" as Helmut Willke notes (1998:164), but knowledge work itself also plays a crucial role in the market-led domains of corporate economies as a concept for sustaining the foundational shift in the "being" of these organisations (Collins 1998, no pagination).[25] In this way, knowledge work contributes primarily to business survival and economic success.

entrepreneurs", these people work in diverse professions and fields of action. The work they do stands for an intangible form of labour frequently dubbed "information manipulation", "programming", "innovating", "facilitating", "creating" or "community founding" etc. (Ibid.:28). Such tags, I suggest, mirror the fuzzy and more ephemeral conditions of cultural production in the post-industrial "knowledge society", while also connoting what these people do and how they act under the conditions of the now global "knowledge society" with its corporate knowledge economies. This means that navigating in "digital information universes"—the time-consuming working engagement of a majority of these people—can be seen as a perpetual fight against ever-growing mountains of information and knowledge. As "on-line curators", "image-entrepreneurs", "net gurus", "knowledge-architects", "music-makers", "fashionistas", "brand-builders" or "dot-commers", as Banks notes (4), cultural workers have thus become the transient "inhabitants" of virtual knowledge-worlds in constant flux and the subjects of the idiosyncrasies of a "surface upon surface" experience, as Roy Ascott notes (2006:109).

[25] This is because knowledge work is linked to organisational knowledge. Moreover, the knowledge of the workforce constitutes an organisation's knowledge base and "intellectual capital", which impacts the value of an enterprise (Kelloway and Barling 2000).

The term "knowledge work" was first popularised by Peter Drucker (1973; 1979), who predicted the growing importance of the "knowledge" worker; since then much research has been conducted into its meaning and relevance for the corporate economy. Yet, numerous authors have struggled with the slippery concept (Collins 1998; Despres and Hiltrop 1995; Drucker 1991; Liu 2004). David Collins, for example, suggests that the concept acts as a brake on academic post-industrial analysis (1998, no pagination). The problem common to analyses of post-industrialism and the knowledge age, he argues, is that the categorisation of work and predictions concerning its future are based on "vague" and "dubious" notions with regard to the "nature of work", the processes of work and the processes of management in general. Furthermore, as Collins outlines, "knowledge work" seems to be the privilege of an elite. Instead of defining and analysing the notion of "work" and "knowledge work", which Collins sees as surrounded by "ambiguity and confusion", he proposes to refocus the debate and analysing what he calls "working knowledge" across the population:

> Yet, we can, if we choose to make the effort, develop and use concepts for quite different ends. Why not, instead of bandying around buzzwords such as knowledge work, begin from the understanding that all workers are knowledge workers and that all have skills and working knowledge, rather than claim it as the possession of a minority group. If we are to regard workers as key resources, why begin from an initial assumption which implies that many have only the most limited resources at their disposal? How much better it would be if, instead of levering buzzwords into currently popular models and ways of thinking about management and organizations, we attempted to develop and apply concepts which could engender different ways

of thinking about problems and innovations within organizations. (Ibid.)

In *The Laws of Cool* (2004), Alan Liu contributes to a more inclusive understanding of the cultural life of information, or, more broadly, of contemporary "knowledge work" in the post-industrial global culture. He theorises knowledge work alongside future perspectives for the humanities and arts, and explores the emergence of new information technologies with potential impacts on the many different forms and practices of knowledge. Furthermore, Banks, drawing on governmentality and liberal-democratic approaches, provides an overview of critical theory approaches in relation to the present and future workplace challenges of the cultural industries. In contrast to Liu's investigation of knowledge work as a class concept, Banks explores the constraints and freedoms of cultural work as the vanguard of a new autonomy which he sees challenged by the hegemony of capitalist social relations. Drawing on a variety of definitions and meanings of the term *knowledge worker,* Liu proposes to keep the definition as broad as possible:

> *Knowledge worker* should now be extended further—even as far as consumers who, for example, use an automated teller machine and thus assume some of the work of clerks and tellers. (...) "Knowledge" thus includes in principle all the varieties of scholarship, research information, advertising, and so on, across the major occupational sectors and on both sides of the "work/leisure" divide. (392)

For Liu "knowledge work"—in the light of New Class theory—is *intellectual* work conducted by academics whose "technical, professional, and managerial function is very much to the point and must be thought positively rather than negatively" (Ibid.). Clerical-level workers, who constitute the "'proletarian' office labor force" (Ibid.) are included in analysis in order to emphasise the study's intention to underpin an ethos of what Liu prospectively calls the "*new* new middle class" (393). Moreover, Liu sees "knowledge work" as an engagement at the "techno-informatic vanishing point of contemporary aesthetics, psychology, morality, politics, spirituality, and everything" and outlines that "beauty, sublimity, tragedy, grace, or evil" are currently being replaced by the new notion of "cool or not cool" (3). In addition, knowledge work, according to Liu, has become a "parallel system of learning—or just as accurately, antilearning" (305). While Liu's study has focused on US knowledge work, it might play a constitutive role on a global scale. According to Liu:

> The contemporary globe, perhaps, is not so much a preexisting object as a standing wavefront of simulation generated by knowledge work, as an *idea* of globalism— named, for example, "new world order", "global market", or "World Wide Web". (287)

In order to circumvent Collins' concept of "working knowledge" a different approach to understanding "knowledge work" and using it as a normative concept is proposed here. In other words, my definition of "knowledge work" (and "cultural work") encompasses both the idea of deeply-ingrained knowledge-practice as well as knowledge-building intentions (cf. Felt, Wynne et al. 2007:59) mediated through *communication*. While communication is not only primarily quantitative, but also qualitatively complex, communication allows "not just for one, but for an increasing number of possibilities of expression and representation", as Gibbons et al. note (1994:43). From such a perspective, then, cultural work can be conceived as a potentially open and constructive form of communication among other existing ways of communicating and expressing concerns or imaginations about the world. In addition, defining and understanding cultural work as a more complex communication/learning process opens it to the notion of a much wider and open panorama for new practices capable of accomplishing inherently (self-)reflexive, socio-political and ecological tasks to produce and disseminate knowledge in the humanities and arts, and science and technology.

## Unsustainable Sociosphere

Although human sustainability implies the integration of social, economic and environmental spheres to meet the needs of the present without compromising the ability of future generations to meet their own needs, the many unsustainable material worlds that we have created, and the severe degradation of the planet's natural ecosystems, mirror our current thinking. Moreover, the treatment of *everything* as a tradable commodity, such as the privatisation of global water resources,[26] shows how ignorance is deliberately produced, and how economic exploitation and unethical pursuits are directly linked to it. What is thus urgently needed today are efforts to live more sustainably, which could take many forms from reorganising living conditions (such as sustainable cities), reappraising economic sectors (such as sustainable agriculture) or work practices (such as sustainable architecture), to adjustments in individual lifestyles, as Goleman notes. According to Goleman, the three interlocking realms of the *geosphere* (soil, water, air and climate), the *biosphere* (our bodies, those of other species, and plant life) and the *sociosphere* (human concerns such as the conditions of workers) (2009:57) need to be "weighed in the equation for improvements" (Ibid.:66). While public reflection on these issues is at stake and reinventing the sociosphere is a very real challenge, democratic knowledge dissemination and collective reasoning are additional key challenges in the struggle for a more sustainable sociosphere in which knowledge practices and cultural work could become more effective.

---

[26] Vandana Shiva notes in *The World on the Edge* (2000): "Energy companies are entering the water sector. General Electric has joined forces with the World Bank (...) to invest billions of dollars in a 'Global Power Fund' to privatise energy and water around the world. Enron has acquired Wessex Water in Britain and is bidding for the $800-billion global water market. Monsanto, the Life Sciences giant, is now leading the race to control water" (125).

## What Knowledge?

The social realities and technoscientific worlds which we inhabit are mirrored in a multitude of knowledges and our ways of thinking about them. But, what knowledge and whose reality should it be? What is the role of the particular realities that the humanities and arts produce if it is no longer possible to make a substantial distinction between the humanities and the sciences in the context of contemporary knowledge production and representation, as Gibbons et al. suggest (1994:103)? Is there a role for "cultural work" to suggest new ways of how reality could be represented and behaviours and relations could be shaped? What is the role of "cultural work" and knowledge in the context of the post-industrial culture in which ecological and ethical issues will become more significant? I believe that these issues are best explored by paying close attention to both cultural production and cultural practices in order to provide a detailed and empirically observed account. Moreover, the complex social role of knowledge and the ways in which knowledge impacts and relates to public opinion are best studied as an element that is more explicitly linked to culturally embedded issues and to communal and institutional assignments or structures (cf. Swidler and Arditi 1994:306). At an epistemological level, my interest lies in the ordinary material-semiotic objects that we create in order to give meaning to our collective and individual lives. The primary focus of my analysis therefore lies on the socio-epistemic and political dimensions of different knowledges and materials including actors with salient concerns, which I will explore in three case studies. The study of the Gotthard Base Tunnel exhibition will provide analyses of material-semiotic responses to issues concerned with the presentation of technoscience to wider publics. Jennifer Baichwal's documentary film *Manufactured Landscapes* will serve to reflect on the film's material-semiotic engagement with and visualisation of the consequences of China's industrial revolution for (global) society at large. The study of Ai Weiwei's multi-million dollar project *Fairytale* at Documenta 12 will explore the aims and motivations underlying the project, and will address the issue of concept-driven conduct, which appears to have become idiosyncratic and hegemonic in contemporary art. Instead of analysing particular patterns of social relations and the labour process itself, I will focus on "cultural content", narratives, and socially and culturally distributed knowledges in particular. As cultural work has been neglected in the literature and is thus significantly under-theorised, as outlined earlier, the questions and issues presented here have also been largely ignored.

# Actor-Network Theory

Actor-Network Theory (ANT) is a distinctive approach to social theory and research, which originated in the field of science studies. Developed within Science and Technology Studies (STS) by the French sociologists Michel Callon and Bruno Latour, the British sociologist John Law and others, it can more technically be described as a material-semiotic method. ANT tries to explain how material-semiotic networks come together to act as a whole. ANT reflects many of the preoccupations of French post-structuralism, and in particular a concern with non-foundational and multiple material-semiotic relations. Its grounding in (predominantly English) STS was reflected in a commitment to the development of theory through empirical case studies, and its links with work on technical systems were reflected in its inclination to extend the analysis of large-scale technological developments to include political, organisational, legal, technical and scientific factors. Among other attempts to empirically describe science and technology, the contributions of ANT anticipated this paradigmatic change in the social sciences (Belliger and Krieger 2006:17). From about 1990 onwards, ANT started to become popular as a tool of analysis in a range of fields beyond STS. It was developed by authors within organisational analysis, computer science, health studies, geography, sociology, anthropology, feminism studies and economics. Today, ANT encompasses a broad and controversial range of material-semiotic approaches for the analysis of heterogeneous relations such as those that I will investigate in this book. ANT is, as the Polish philosopher Ewa Bińczyk suggests, a useful instrument to diagnose the "global, uncertain, changeable, contemporary conditions of reality and society", as the theory provides a "new epistemological standpoint that does not privilege ontology over epistemology" (2008:205).

While ANT's *principle of generalised symmetry,* to which I refer below, circumvents the unequal status of ontology and epistemology in Western philosophy, traditional *analytic epistemology* does not question the Western paradigm of knowing, which can be exemplified by the Cartesian individual or *knower* (Fuller 1988; 2002:ix). As a result, the Cartesian perspective has sustained anthropocentric views and perspectives on knowledge for a long time.[27] Thus, a conceptual challenge of my work is to view cultural work and knowledge from an (ideally) non-dualist perspective that treats the domains of ontology and epistemology equally—an unsolved problem for the sociology of knowledge's strong programme.[28]

I follow Latour's prescription for an anthropology of science to a large extent, and will adapt it to the contemporary cultural context in which the agency of non-humans, material-semiotic objects and artefacts play roles in defining the public interest in different domains.[29] I am inspired by Latour's originality in offering a distinct non-dualising philosophy in which neither nature nor society, neither the objective nor the subjective, are taken for granted, and which can even problematise the agency of matter, as Bińczyk outlines (2008:206; Latour 1999a:125), by incorporating it into empirical case studies (Ibid.:205). I should add that Latour's criticism of the modern constitution and the closing of our eyes to the hybrid-

---

[27] The theoretical frameworks of the Sociology of Scientific Knowledge (SSK) have received criticism from the main theorists in the ANT school, Michel Callon, Bruno Latour and John Law. SSK has been criticised for sociological reductionism and a human-centred universe. SSK is said to rely too heavily on human actors, social rules and conventions in settling scientific controversies. The ANT school, instead, proposes that non-human actors (actants) play an integral role. For example, instruments, measurement scales, laboratories etc. have the unintentional capacity to end a scientific controversy.
[28] The strong programme of the sociology of knowledge formulated and supported mainly by two representatives of the Edinburgh School, Barry Barnes and David Bloor, exhibits what Bińczyk refers to as its "tiny weakness", namely the dualistic opposition between ontological and epistemological domains (2008:203).
[29] While human and non-human actors such as artefacts, institutions, material-semiotic structures, or other objectified norms or things are integrated into the same conceptual framework and are thus given equal status of agency, the resultant "methodological symmetry" allows research conducted in both science and technology, and the domain of theories of human society to be treated equally (Belliger and Krieger 2006:15).

ity of the machines and the "monsters" that are thus produced (Latour 1993:131)—the tyranny of social interest dominating us—has supported me in developing a critical view on issues such as economic rationality, technological efficiency and scientific truth.

The research presented in this book is confronted with some limitations in scope. My "fieldwork" and theoretical analysis have given me many insights into the methodologies, theories and hybrid practices of cultural work, including its significance and value. Yet some of the limitations of my perspective can be seen, for example, in the fact that I do not discuss how each case study was financed. My work, then, is not always based on a complete network analysis. A further intellectual challenge is the proper treatment of what Actor-Network Theory attempts to overcome and what its proponents (among them Latour in particular) consider to be *the* major problem of Modernism and Postmodernism: the cutting up of a "hybrid" reality into analytical realms. While ANT's key feature—the distinction between humans and non-humans—is intriguing, the conventional sociological notion of actors as "social entities" is not entirely given up in my work. Throughout my text the aspect of action and "doing things" is emphasised, yet the notion of what actors ("cultural workers") *are* (moral "entities", for example) will also be considered.

I wish to state that it would have taken a major research programme— far beyond my resources—to collect the necessary data in order to establish the exact limits for many of my arguments. In the next chapter, I will initially venture into Actor-Network Theory as a (socio-)material-semiotic and methodological tool in relation to the three case studies and clarify some of the misunderstandings surrounding it. I will then introduce the case studies from the perspective of Actor-Network Theory and material semiotics.

# II – Cultural Spaces of Knowledge and "Cultural Knowledge Work"

Bakhtin's concept requires us to enter the contingency, thickness, inequality, incommensurability, and dynamism of cultural systems of reference through which people enroll each other in their realities. (Haraway [1997a:42] referring to Bakhtin's [1981] concept of the chronotope as a figure that organises temporality)

If we want to do philosophy, metaphysics and politics, and if we want to explore both the character and the quality of knowledge, we cannot proceed in the abstract. We need to proceed empirically (Law 2009b:2). This means, as Law points out, that in order to understand how realities "are done" or to explore their politics, we have to look at practices and ask how these practices work (Ibid.). While the sociologists of science worked through exemplary case studies, the nascent actor-network writers, also within the sociology of science and technology, have followed an identical path in order to shed light on the knowledge and the practices which lie in exemplars, since "words are never enough" (Law 2009a:144). Thus, for the purpose of studying cultural work practices, different enacted realities and forms of knowledge, I use the actor-network approach in the three case studies to detect the "somewhat ordered sets of material-semiotic relations", in order to undertake what Law defines as the "analytical and empirical task of exploring possible patterns of relations" (2009b:1). Not only are these relations materially and semiotically assembled in particular places and spaces—the public domain of the Gotthard Base Tunnel exhibition, for example—but whatever it is that is being assembled—such as materials, objects, images, people, meanings, realities and knowledges—depends on practices that constitute assemblages of relations in a particular moment, location, occasion or medium, of which *Manufactured Landscapes* is an example. Thus, the material-semiotic entity of an exhibition, a film such as *Manufactured Landscapes*, or the socially and materially heterogeneous system of *Fairytale* at Documenta 12 are what Law calls *assemblages,* which generate specific realities or "incidental collateral realities [which are] inseparable from the patterning juxtapositions of practices" (Ibid.:2).

The (socio-)material-semiotic approach applied in this book is based on Law's definition of ANT and is one form of "material semiotics" among other possible forms. It has helped to define the analytical perspective underlying the three case studies in order to approach the empirical material in more descriptive rather than explanatory ways. The key issue is to explore "stories" about "how" relations assemble and "how" social and cultural knowledge spaces are manufactured in order to develop a sensibility towards what Law calls the "messy practices of relationality and materiality of the world" (2009a:141). Hence:

> Actor network theory is a disparate family of material-semiotic tools, sensibilities, and methods of analysis that treat everything in the social and natural worlds as a continuously generated effect of the webs of relations within which they are located. It assumes that nothing has reality or form outside the enactment of those relations. Its studies explore and characterize the webs and the practices that carry them. Like other material-semiotic approaches, the actor network approach thus describes the enactment of materially and discursively heterogeneous relations that produce and reshuffle all kinds of actors including objects, subjects, human beings, machines, animals, "nature",

ideas, organizations, inequalities, scale and sizes, and geographical arrangements. (Ibid.)

As I will rely on Actor-Network Theory as a methodological tool for the analysis of the three case studies, the term "network" requires some clarification. The word "network" normally implies more technical dimensions such as a telephone "network" or a computer "network". However, according to Latour, an "actor network" is a different entity:

> The first mistake would be to give it a common technical meaning in the sense of a sewage, or train, or subway, or telephone "network". Recent technologies have often the character of a network, that is, of exclusively related yet very distant element [sic] with the circulation between nodes being made compulsory through a set of rigorous paths giving to a few nodes a strategic character. Nothing is more intensely connected, more distant, more compulsory and more strategically organized than a computer network. Such is not however the basic metaphor of an actor-network. A technical network in the engineer's sense is only one of the possible final and stabilized state [sic] of an actor-network. An actor-network may lack all the characteristics of a technical network—it may be local, it may have no compulsory paths, no strategically positioned nodes. (1999b:1)

Thus, Actor-Network Theory has very little to do with the study of *social networks*. In Latour's view (Ibid.), it aims to account for the "very essence of societies and natures" since it contributes more to ontology and metaphysics than to sociology. While Actor-Network Theory was devised to rethink more globally functioning entities or structures such as institutions, organisations and even states in an attempt to describe the very nature of societies as well as the agency of non-humans (machines, animals, texts etc.), the ANT network is conceived as a "network" of associations between *volitional* actors, termed *actants,* that can associate or disassociate with other non-human agents in the process of creating realities.[30] From such a perspective, the "actors" within the three case studies can also be conceived as an amalgamation of symbolically invested "things", "identities" and "meanings", which are expressed in diverse materials that are embedded in multidimensional "networks" and the reflective practices of human actors.

In an attempt to explore the social, cultural and political dimensions of the three case studies, a key issue is the socio-epistemic perspective of cultural work as well as the question of how to redesign processes of cultural and scientific learning in order to envisage alternative modes of the governance of science and technology. While "actor-networks" can be seen as scaled-down versions of Foucault's epochal epistemes and discourses, as Law suggests (2009b:145), the analysis of different "actors" in the case studies (over potential others) contributes to the perspective of the Foucauldian conception of power and knowledge. And whereas the case studies investigate the agency of non-human/human actors and the knowledge spaces constructed by them, the material-semiotic properties of the non-human actors, such as the "objects" in the Gotthard Base Tunnel exhibition or the film stills of *Manufactured Landscapes*, are equally important. Yet, instead of analysing the entire empirical material (the entire collection of objects in the exhibition and the film's complete narrative), I have chosen material-

---

[30] According to Law, in actor-network webs the distinction between human and non-human is of little initial analytical importance as "people are relational effects that include both the human and the nonhuman". "An actor is always a network of elements," he outlines, "that it does not fully recognize or know: simplification or 'black boxing' is a necessary part of agency" (2009a:147).

semiotic "objects" and film stills with a specific social, epistemic or technical quality. In my analysis, I will attempt to describe and reflect on these objects in order to expand my theoretical understanding of both knowledge practice and cultural work practice. The number of objects chosen for analysis is of secondary significance. What is relevant is the potential of these objects to tell "stories" and to provide insights into the field being studied (Taylor and Bogdan 1998:93). Each of the case studies will thus provide analyses of the (socio-)material-semiotic, conceptual and heterogeneous dimensions of cultural work in the post-industrial culture.


## "Material Semiotics"

Actor-Network Theory as a diaspora (Law 2009a:142) overlaps with other intellectual traditions. "Material semiotics" is perhaps the better term for it as it entails the "openness, uncertainty, revisability, and diversity of the most interesting work", as Law claims, and is not a "creed or a dogma", but "at its best" a degree of "[intellectual] humility" (Ibid.). With a focus on this "material semiotics", the case studies will shed light on the purposes and interests behind cultural work and knowledge practices, namely:

- "how" the Gotthard Base Tunnel exhibition as a constructed public techno-socio-cultural space of knowledge and the result of cultural work depends on deeply embedded knowledge practices amidst wider economic and political convergences.
- "how" *Manufactured Landscapes* functions as a moral-political space of knowledge for public reflection and the outcome of cultural work.
- "how" the *Fairytale* project as a socio-epistemic, political and "dialogic" space amidst the wider global market culture is rooted in and depends on practices of different cultural and economic actors.

The three studies do not intend to present an "overall view" in terms of an apparent claim to objectivity. The principal aim is instead to reflect on cultural work practice and material-semiotic outcomes, which lead to diverse narratives pertaining to particular causes and interests of different actors trying to build meaning in society. While the heterogeneity of the three case studies is significant in itself, the actor-network approach is applied in each study in slightly different ways. The ANT approach will help to explore what Law has described as the "strategic, relational, and productive character of particular, smaller-scale, heterogeneous actor networks" (2009a:145). Furthermore, ANT will help to address what different actors do, and how they act under contingent conditions.

In each case study, I will reflect on how Actor-Network Theory has served as a conceptual and methodological tool (including its limits) in exploring cultural work practices. I will also briefly address the major strands of criticism of ANT (Walsham 1997). In the case study of the Gotthard Base Tunnel exhibition I will articulate the nature of Actor-Network Theory's problems with description, and in *Manufactured Landscapes* I will elaborate on ANT's criticised stance on moral and political issues. Finally, in the case study of *Fairytale,* I will discuss the criticism that ANT addresses the local and contingent, but pays little attention to wider social structures. I will then offer personal views of the relevance of these criticisms for my research.

## Knowledge Production and Dissemination

Cultural work encompasses a broad range of diverse interests based on behaviours or purposes that are often "programmatic", "innovative", "facilitative" or even "manipulative", and which expose the problematic and contingent character of current practices. Yet, cultural work seems at the same time to have become more of what we may call an *intangible* activity and engagement (as outlined in footnote 24 in the Introduction). If my analysis is correct, then cultural work encompasses an as yet unexplored contingency and incommensurability of knowledge production and knowledge dissemination in the humanities and the arts, and science and technology. For these knowledge-based and potentially reflective capacities of cultural work, I have created the term "cultural knowledge work" (CKW) (Figure 2.1). The conceptual and practical challenge is, as outlined earlier, to explore cultural work as a form of human communication and "knowledge-making", of which the issue of the wider societal distribution of knowledge is particularly challenging.

A key issue for an "epistemology" of cultural work under present conditions is thus the development of democratic capacities in order to create new opportunities where contingencies can be addressed (cf. Felt, Wynne et al. 2007:69). As outlined above, another salient issue for cultural work that must be recognised as a challenge in itself lies in the larger context of the newly dominant corporate knowledge cultures and the discourse of "enterprise" that the cultural industries have assimilated (cf. Banks 2007:42). Furthermore, it follows from Liu's work that the humanities and arts are subject to the pressure of these globe-spanning knowledge cultures and learning organisations etc. The crucial question is whether in the age of corporate knowledge work and the ruling rationalities of a global market culture, cultural work will be permitted to obtain a foothold in a "knowledge space" that is different from the one it finds itself embedded in the present organisations and activities of the cultural industries. Another important question is whether there are "alternatives" in the cultural production of knowledge that might provide significant and "reflexive" counterpoints to the unsustainable and unecological material worlds that we have created.

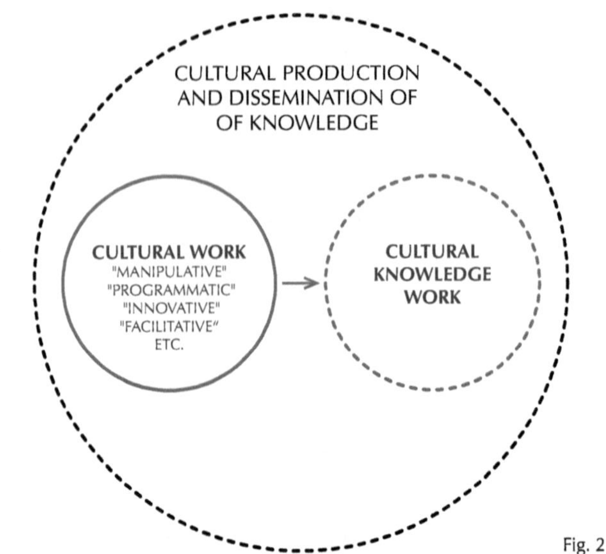


Fig. 2.1

In the next chapter, I will provide an in-depth analysis of the Gotthard Base Tunnel exhibition and address issues in relation to the normative, economic, political as well as epistemological dimensions and conditions to which the construction of the tunnel is linked. The key question is that of the cultural representation of *technoscience*. This includes public reflection on the issue, as the exhibition provides numerous material-semiotic responses to technoscientific knowledge and progress, and employs a range of educational formats to stimulate public discourse.

## Building a Transalpine Railway Tunnel and Manufacturing a Public and Transient Cultural Space of Knowledge

> The Navy's organization is profoundly modified by the way its offices are allied with its bombs; EDF and Renault take on a completely different look depending on whether they invest in fuel cells or the internal combustion engine; America before electricity and America after are two different places; the social context of the nineteenth century is altered according to whether it is made up of wretched souls or poor people infected by microbes; as for the unconscious subjects stretched out on the analyst's couch, we picture them differently depending on whether their dry brain is discharging neuro-transmitters or their moist brain is secreting hormones. (Latour 1993:4)

In this case study, I venture into an assessment of the diverse social, economic, political, epistemological and "networked" characteristics of cultural work. In thus looking more closely at the "networks" in which cultural work is embedded and in recognising the hidden dimensions of the "networked whole" of human and non-human actors, I thus open up the very substance of society for exploration and deliberation. Yet, while Latour's examples above invoke both this substance and an immaterial stage on which knowledge, reflexivity, politics, production and timely actions are performed, the networks themselves are not visible as they resemble invisible spiders' webs (1993:4).[31] As a consequence, my aim is to understand such clandestine and complex networks while elaborating

---

[31] According to Manuel Castells, the post-industrial world, the economy, firms and their territories are organised in networks of production, management and distribution whose core economic activities are global, with a capacity to work "as a unit in real time, or chosen time, on a planetary scale" (2000:52).

on my exploration of "cultural work" as a practice that is tacitly shaped and framed by deeper cultural forces, values and interests. Overall, in this study I will follow more descriptive and analytical approaches.

Fig. 2.2

In dealing with tunnel construction, the Gotthard Base Tunnel exhibition seeks to provide knowledge and information about technological and science-based issues such as the state-of-the-art technology of new computer-controlled tunnel boring machinery, satellite systems to map the tunnel, safety matters on the construction site including the exhibition's implicit concern to promote faster train services, better connections and quicker journeys, and other issues connected with risk management as well as the environment. I will rely only on a small portion of "facts and figures" presented in the exhibition that informs the visitors as to how the tunnel is constructed and under which future conditions rail transportation and freight traffic across the Alps will work. My analysis is based on more philosophical premises and intentions. I will primarily reflect on how public awareness is created and how this is related to issues such as the exhibition's main concern of serving as a catalyst for knowledge exchange and "civic dialogue" in relation to the larger discourse on technoscience as a source of progress and scientific learning.

I have chosen the "InfoCenter" near the village of Erstfeld for my study (two further centres were built at Pollegio and Sedrun).[32] I took the photograph above during the winter of 2008 and 2009 (Figure 2.2).[33] The exhibition centre is located in the flat red building, which can easily be seen from afar when approaching the construction site. A typical construction used on industrial and laboratory research sites around the globe, the structure first reminded me of the famous World War II huts at Bletchley Park.[34] The installations in the background that resemble a roller coaster are conveyor belts. These belts are part of the site's large surface installations and pipelines that provide the water supply and electricity, and that will transport an estimated 24 million tonnes of excavated rock to the surface.[35] From March to December 2008 about 20,000 visitors came to see the exhibition.[36] By comparison, the Museum of Art Lucerne, the

---

[32] In 2001, the cantonal governments of Uri and Ticino called for architectural proposals for the two planned exhibition centres at Erstfeld and Pollegio. Twenty-seven designs were submitted. The winning project was built between 2001 and 2003 at Pollegio in the Italian-speaking part of Switzerland using excavated rock from the tunnel. However, constructing a second centre in Erstfeld became a "politically problematic" issue, according to Ambros Zgraggen, the media spokesman of AlpTransit Gotthard Ltd. The new plans were again based on the idea of building an exhibition centre with a "modern" and "emotional" ambience by combining given features such as "permanence" (the tunnel) and "impermanence" (information and communication technology). Public debate on the issue started when the exorbitant tunnel costs incurred by the government of Uri and the Swiss government became a topic in the media. Additionally, when Swiss taxpayers started to criticise the planned expenditure, the high costs of 10 million Swiss francs previously spent on the centre in Pollegio caused AlpTransit Gotthard Ltd. to resize the exhibition project at Erstfeld. The Erstfeld information centre was finally realised in a modest hut, yet with costs still amounting to 1 million Swiss francs. Unterschütz 2002a; <http://www.bauzeit.com/content/public/detail.php?cat=2&subcat=3&ID=59> downloaded 4 February 2009; and interview between the author and Ambros Zgraggen, media spokesman, AlpTransit Gotthard Ltd., which took place on 20 January 2009.
[33] The Erstfeld section is the most northerly section of the tunnel. The first part of this section was constructed by digging an open trench, which was covered over after completion. The remainder was cut with tunnel boring machines. *The New Gotthard Rail Link*. AlpTransit Gotthard Ltd. (Ed.), 2005:18.
[34] Bletchley Park, National Codes Centre, Milton Keynes, England, was the location of the UK's main codebreaking establishment during World War II.
[35] *The New Gotthard Rail Link*. AlpTransit Gotthard Ltd. (Ed.), 2005:30.
[36] *Über 20'000 Besucher im InfoCenter Erstfeld*. Press release (in German), AlpTransit Gotthard Ltd., 19 December 2008.

largest art institution in central Switzerland, had 46,000 visitors in the same period.[37]

## Historical, Political and Economic Significance

Initially, I focus on the historical, political and economic significance of the construction of the tunnel, whose estimated costs amount to 12 billion Swiss francs.[38] The idea of building a Gotthard Railway Base Tunnel is not new. The first such proposal was put forward as early as 1947. The first project was initiated by the Swiss Federal Department of Home Affairs in 1962. The acceptance of the proposals for the New Rail Link through the Alps (NRLA) in 1992, as part of the creation of a pan-European network of high-speed railways, provided the basis for planning (Figure 2.3 *The European High-Speed Rail Network in 2020*).[39]

The acceptance of the Heavy Vehicle Tax (HVT) in 1998, as well as the proposal to modernise the railways, finally cleared the way for construction. The history of the Gotthard is the history of transportation, road and tunnel

[37] Personal mail dated 22 January 2009 from Doris Bucher, Head of Communications, Museum of Art Lucerne. Shortly before and after the opening, the exhibition began to attract the attention of the public and the media. Groups of visitors and large audiences came to see it. However, the media interest declined once the exhibition had opened its doors. Twenty-six review articles were published in local newspapers such as the *Neue Luzerner Zeitung* or the *Neue Urner Zeitung* between 15 March 2008 and 2 June 2010. The reports informed the public about the location, exhibition content and public response including marketing efforts to promote the canton of Uri as a tourist asset. Most articles mentioned the tunnel boring machinery or other eye-catching elements such as a film about blasting and a statue of William Tell. The required technical apparatus and the tunnel's support system, tunnel history, finances and politics were other topics covered the media.

[38] Financing comes mainly from a kilometre-based tax on heavy goods vehicles, as well as from part of the taxes on fuel originally intended for road building, and to a lesser degree from value added tax funds. The figure was given by Ambros Zgraggen, media spokesman, AlpTransit Gotthard Ltd., on 20 January 2009 in Lucerne. In the meantime figures available on the internet amount to 16 billion Swiss francs. <http://en.wikipedia.org/wiki/AlpTransit> downloaded 20 April 2013.

[39] *The New Gotthard Rail Link*. AlpTransit Gotthard Ltd. (Ed.), 2005:2. The new Gotthard Rail Link constitutes an essential backbone of the *Trans-European-Network-Project* Number 24 (freight corridor) connecting Rotterdam to Genoa as an important part of the pan-European network of high-speed railways (Kallas and Truttmann 2010:3).

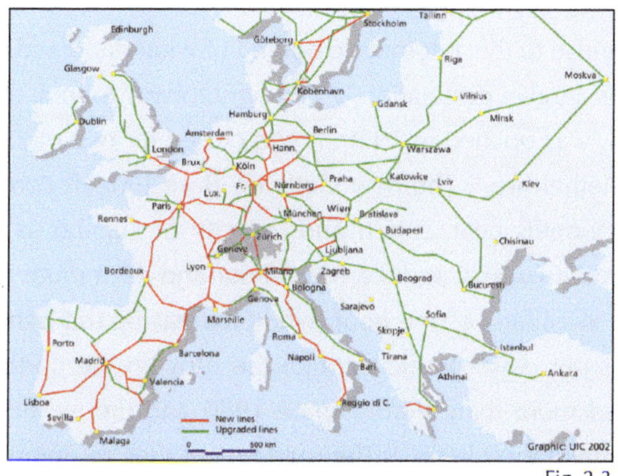


Fig. 2.3

construction. This is because of its geographical importance as the shortest road link from Hamburg, Germany, to Sicily in Italy. The first route over the Gotthard Pass opened around 1220, resulting in economic growth due to the movement of merchants and travellers along one of the most important passages through the Alps on the north-south axis in Europe.[40] Today, several tunnels provide access through the pass. The 15km Gotthard Rail Tunnel was built between 1872 and 1882 at a cost of hundreds of workers' lives. It replaced the road pass. Its construction was difficult due to financial, technical and geological issues, the latter leading to the death of workers mainly due to water inrushes. Many workers were also killed by the compressed air-driven trains that transported the excavated material out of the tunnel.[41] The poor working and living conditions led to a violent strike by workers in 1875.[42]

[40] The St. Gotthard Pass (Italian: San Gottardo, 2108m above sea level) is a high mountain pass between the village of Airolo in the canton of Ticino, and the village of Göschenen in the canton of Uri, connecting the northern German-speaking part of Switzerland with the Italian-speaking part along the route onwards to Milan. As early as 1237, it was dedicated to the Bavarian Saint Gotthard of Hildesheim (Italian: Monte Sancti Gutardi). Mäder 2002:179-97; and *Historisches Lexikon der Schweiz* (HLS) <http://www.hls-dhs-dss.ch/textes/d/D7466.php> downloaded 3 July 2009.

[41] *Historisches Lexikon der Schweiz (HLS)* <http://www.hls-dhs-dss.ch/textes/d/D7962.php>; Alp-Transit Gotthard AG Home: <http://www.alptransit.ch/fileadmin/documents/PDF/geschichte.pdf> downloaded 15 October 2010.

[42] Schnieper (2002:147); and *Historisches Lexikon der Schweiz (HLS)* <http://www.hls-dhs-dss.ch/textes/ d/D16386-3-6.php> downloaded 3 July 2009.

As a response to the automobile boom in Switzerland and the popularity of Italy as a travel destination, a 17km motorway tunnel, the Gotthard Road Tunnel, was opened in 1980. It was closed for two months in 2001 following a lethal fire.[43] The new Gotthard Base Tunnel, combined with two shorter tunnels near Zurich and the city of Lugano as part of the AlpTransit initiative, will reduce the 3-hour-and-40-minute rail journey from Zurich to Milan by one hour, while increasing the transit capacity to 300 trains per day.[44] Traffic through the Alps on the north-south axis has increased more than tenfold since 1980 and the existing road and rail tunnels are at their limits.[45] The road tunnel sees an immense amount of traffic and is often subject to traffic jams at both the north and south ends. In order to ensure a faster passage through the Alps, Swiss voters have decided to build a new tunnel cutting through the Gotthard massif at nearly ground level, 600 metres below the existing railway tunnel. On the current track only certain freight trains with a maximum weight of 2,000 tons are able to pass through the narrow mountain valleys and through spiral tunnels climbing up to the portals of the old tunnel at a height of 1,100 metres above sea level. Once the new tunnel is completed in 2016, standard freight trains of up to 4,000 tons will be able to pass the Alps.[46]

Because of the ever increasing international large goods vehicle (LGV) traffic, the Swiss voted in 1994 for a shift in transportation policy (Traffic Transfer Act of 1999).[47] The goal of the law and the Gotthard Base Tunnel

[43] <http://en.wikipedia.org/wiki/Gotthard_Road_Tunnel> downloaded 4 July 2009.
[44] Steinmann, N., Favre, P. (2003). *Building a Modern Railway Line in the Gotthard Base Tunnel*. Lucerne: AlpTransit Gotthard AG.
[45] <http://en.wikipedia.org/wiki/Gotthard_Base_Tunnel> downloaded 14 January 2009. Until the opening of the Gotthard Road Tunnel in 1980, the Swiss Federal Railways offered piggyback services for cars and LGVs through the Gotthard Tunnel. Today, that service exists as the Rolling Highway from the German to the Italian border and aims to reduce LGV traffic on Swiss motorways. <http://en.wikipedia.org/wiki/Rolling_highway> downloaded 3 July 2009.
[46] *The New Gotthard Rail Link*. AlpTransit Gotthard Ltd. (Ed.), 2005:10.
[47] <http://www.bav.admin.ch/glossar/index.html?lang=en&action=id&id=110> downloaded 14 January 2009.

is to transport LGVs, trailers and freight containers from southern Germany to northern Italy in order to relieve the overused roads, and to meet the political requirement of shifting as much tonnage as possible from LGV transport to train transport (Alpine Protection Act of 1994; German: "Alpen-Initiative").[48] Efforts to build a second tunnel have failed, blocked by political resistance, but are currently being reinvigorated.[49] The Alpine Initiative for the protection of the Alpine region from transit traffic, which raised barriers against road tunnel construction, was initially blocked by the Swiss Parliament. However, in 1994 it was passed by 52% of voters, and Parliament upheld the referendum twice throughout the 1990s. The pro-tunnel Avanti Initiative brought a referendum to voters in 2004, which was rejected by 62.8% of the electorate.[50] As we can see in these complex historical and political dimensions surrounding the building of a new railway tunnel across the Alps, history and politics are anything but "only" events. They have deeply ingrained social and performative dimensions interwoven into them, and they involve different actors, publics and interests.

## "Anthropological Field Trip"

My concept consists in reflecting on the knowledge and the information presented in the exhibition. I have selected a total of 26 display cases, info-terminals and video screens. Being something of a "cultural epistemologist", I follow the displays and will primarily reflect on the material-semiotic properties of these "objects". In contrast to Latour's field trip in the Amazon, where he conducted the sampling of soil under laboratory

---

[48] <http://www.alpeninitiative.ch/e/ProtectionOfTheAlps.asp> downloaded 14 January 2009.
[49] <http://www.gotthardtunnel.ch> downloaded 3 July 2009.
[50] <http://en.wikipedia.org/wiki/Gotthard_Road_Tunnel> downloaded 14 January 2009.

conditions, I will embark on an "anthropological field trip" across a physical and built space of knowledge. I will not be permitted to take "specimens" of the exhibited images, texts, models and videos like a biologist, who would collect them in nature, or a pedologist, who digs holes in the ground in order to take soil samples. I will "only" engage in taking a look at what is presented, take pictures with my camera and collect "information" on paper. I will follow the "cultural entities" that evolve before my eyes as something of a "preconstructed universe"—a metaphor which Bruno Latour and Steve Woolgar have used for the laboratory in science (1986:236). My "field trip" into the heart of the Swiss Alps to the Gotthard Base Tunnel exhibition thus requires only little knowledge. I rely on Latour's prescription for an anthropology of science, which I will adapt to the exhibition (the "cultural context"), in which forms of technoscientific knowledge as well as social, economic, political and epistemological priorities and purposes are solidified.

I start with a prominently placed first display in the hallway just next to the main entrance (Figure 2.4). It contains a small statue of a popular woman venerated by those whose life is threatened by violent death at work. The sculpture is of Saint Barbara, who is often depicted standing by a tower.[51] Here she holds a miniature tower in her left arm. Saint Barbara is best known as the patron saint of artillerymen, armourers, military engineers, gunsmiths, miners and anyone else who works with cannons and explosives because of her legend's association with lightning. I am consumed by such materialised representation. I try to understand what is

---

[51] According to the legend, she was guarded by her father who kept her locked in a tower in order to preserve her from the outside world and from becoming a Christian. During her father's absence Barbara had three windows put in the tower, as a symbol of the Holy Trinity. When her father returned, she confessed to being a Christian, whereupon he decided to kill her. However, her prayers created an opening in the tower wall, allowing her to escape. Pursued by her father and guards, Barbara hid in a gorge in the mountains and stayed there until a shepherd betrayed her. As legend has it,

Fig. 2.4

being articulated in material-semiotic terms, entangled in what Haraway regards as "different kind of knots" (1997a:23)—the physical depiction of shared beliefs and values.[52] While the artefactual representation of the female martyr, whose real existence is doubtful, mirrors authority and attracts people, other salient dimensions deeply ingrained in this object such as the transcendental foundations and forms of truth, as well as the problems involved therein, remain beyond articulation.[53] Although not irrelevant they challenge us to think about essential ontological uncertainties and ambiguities that we all share.

the shepherd was transformed into a marble statue and his herd into grasshoppers. Unterschütz 2002b:157-59; and <http://en.wikipedia.org/wiki/Saint_Barbara> downloaded 5 January 2009.
[52] In pointing to the problem of representation, Haraway outlines the astounding *power* of narratives. They can turn, she notes, into "clear mirrors, fully magical mirrors, without once appealing to the transcendental, or the magical" (1997:24).
[53] Although the legend of Saint Barbara is included in the *Golden Legend* and in William Caxton's version of it, and although she was one of the most popular saints of the Middle Ages, some scholars doubt the legend's veracity and even her existence. Because of these doubts about the historicity of her legend, she was removed from the official Catholic calendar in 1969. Unterschütz 2002b:157-59; and <http://en.wikipedia.org/wiki/Saint_Barbara> downloaded 5 January 2009.

Upon entering the main hall we encounter William Tell, a glass-fibre-reinforced polyester version of the original located on the main square in the nearby village of Altdorf.[54] Tell is depicted as a peasant and man of the mountains with strong features and muscular limbs—a legendary hero of disputed historical authenticity again, who is said to have lived in the canton of Uri in the early 14th century, and to have been punished by being forced to shoot an apple off the head of his son Walter.[55] The original shows Tell's hand resting on the shoulder of his son. This depiction is in marked contrast to that used by the Helvetic Republic, in which Tell is shown as a "landsknecht"[56] rather than a peasant, with a sword at his belt and wearing a feathered hat, bending down to pick up his son, who is still holding the apple.[57]

Visitors are invited to stand beside Tell, the supposed hero, and to pretend to take the place of Walter, his son (Figures 2.5).[58] Such a material-semiotic and disneyfied representation of William Tell as a problematic identification figure raises questions with regard to the curatorial concept, which aims to produce and convey particular cultural meanings based on

[54] The original was created by the Swiss sculptor Richard Kissling in 1895.
[55] According to the legend, William Tell was known as an expert marksman with the crossbow. At the time, the Habsburg emperors were seeking to dominate Uri. Hermann Gessler, the newly appointed Austrian *Vogt* of Altdorf (an overlord over ecclesiastical institutions and their territory), raised a pole in the village's central square with his hat on top and demanded that all the local townsfolk bow before it. As Tell passed by without bowing, he was arrested. He received the punishment of being forced to shoot an apple off the head of his son Walter, or else both would be executed. <http://en.wikipedia.org/wiki/William_Tell>; and *Historisches Lexikon der Schweiz* (HLS) <http://www.hls-dhs-dss.ch/textes/d/D17474.php> downloaded 11 January 2009.
[56] A *landsknecht* was a European and most often German mercenary pikeman. <http://en.wikipedia.org/wiki/ Landsknecht> downloaded 5 January 2009.
[57] <http://en.wikipedia.org/wiki/William_Tell> downloaded 11 January 2009.
[58] In 1891 Wilhelm Öchsli published a scientific account of the founding of the confederacy that was commissioned by the Swiss government for the celebration of the first Swiss national holiday on 1 August 1891. It dismissed the William Tell story as fiction. Yet, 50 years later in 1941, when Tell had again become a national identification figure, the historian Karl Meyer tried to connect the events of the saga with known places and events. Modern historians generally consider the story to be fiction, as neither Tell's nor Gessler's existence can be proven. According to a survey, however, 60% of the Swiss believe that Tell really lived. Historians continued to argue over the issue until the 20th century. <http://en.wikipedia.org/wiki/William_Tell>; and *Historisches Lexikon der Schweiz* (HLS) <http://www.hls-dhs-dss.ch/textes/d/D17474.php> downloaded 11 January 2009.

Fig. 2.5

deeper social values and interests. The cultural system of the exhibition thus requires us to pose deeper questions with regard to the incommensurability and contingency on which the concept is based and through which its makers enrol us in multicoded and fuzzy realities—the outcome of deliberate "reflexive" cultural work practice. Just next to William Tell another large display informs visitors about the canton of Uri in order to promote this region in the heart of Switzerland as a tourist destination with white mountain peaks, glaciers, and clean rivers.

As I move on, I stand now right in front of the next display, a massive video screen with nine monitors on which short portrayals of the miners and engineers, 27 in total, are presented (Figure 2.6), a tiny portion of the entire workforce of 2,500 tunnel workers employed on the five construction sites.[59] While these 27 workers and their testimonies constitute an entertaining physical account of the human capital,[60] presenting them as both professionals and ordinary people with family relationships and pri-

[59] The figure was given by Ambros Zgraggen, media spokesman, AlpTransit Gotthard AG, on the occasion of the interview, which took place on 20 January 2009 in Lucerne.
[60] I refer here to Adam Smith, the Scottish philosopher (1723–1790). For Smith, human capital was the stock of skills and knowledge embodied in the ability to perform labour in order to produce economic value.

vate concerns such as the salaries they earn generates a more ambivalent narrative. Much of what we might want to learn in more detail about these individuals and their apparently bold lives remains again unarticulated and beyond reach. Finally, a machine and technoscientific monster constitutes the very heart of the Gotthard Base Tunnel exhibition offering an awe-inspiring grasp of the power needed to master geological uncertainties,

Fig. 2.6

including the numerous incalculable risks of drilling and blasting.[61] The giant is a so-called gripper tunnel boring machine.[62] A full-scale reconstruction of its control cabin and a much smaller model provide an idea of this superstructure measuring 440 metres in length. Yet, compared to the cockpit of a modern passenger airplane, the cabin and its data screens seem rather unspectacular (Figures 2.7 and 2.8). But, it is exactly this not particularly interesting cabin that shapes our imagination as we see ourselves actively steering the earthworm deep inside the mountain in real-time, floodlit in bright artificial

---

[61] *The New Gotthard Rail Link.* AlpTransit Gotthard Ltd. (Ed.), 2005:16, 28.
[62] Tunnel boring machines are used as an alternative to drilling and blasting methods in rock and conventional "hand mining" in soil. In the US, the first boring machine to have been built was used in 1853 during the construction of the Hoosac Tunnel (Western Massachusetts). See <http://en.wikipedia.org/wiki/Tunnel_boring_ machine> downloaded 19 January 2009.

light. We may perhaps even go further and press one of its electronic but-
tons to make the huge grinding head of 19 metres in diameter rotate. Or we
might simply imagine the energy consumption of 2,500 cooker hotplates (if
we can), or the 50,000 light bulbs equalling the electric power needed to run
the technoscientific beast.[63] Yet, to link all this to what is *really* going on inside
the mountain—real explosions, real drilling noise, real dust, and the heat of

Fig. 2.7

45 °C cooled down to 28 °C by water circulating in pipes—remains beyond
our imaginary reach. Detonations filmed in slow motion and presented on a
large video screen offer only a vague notion of the realities which the tunnel
workforce deep inside the Gotthard face in the process of building the longest
tunnel on Earth (Figure 2.9).

---

63 *The New Gotthard Rail Link*. AlpTransit Gotthard Ltd. (Ed.), 2005:25.

Fig. 2.8

Fig. 2.9

## Conclusions

The Gotthard Base Tunnel exhibition constitutes a cultural response by political, economic and institutional stakeholders. The public is invited to follow a public exhibition programme constructed by these actors and based on socio-economic-political priorities and juxtaposed materials such as computer terminals with information on the tunnel construction, technoscientific frontiers (the model of a tunnel boring machine, for example), interactive relief maps promoting tourism as well

Fig. 2.10

as diverse other heterogeneous elements (Saint Barbara and William Tell, for example). By seeing cultural work as a form of engagement which is dependent on realities and relations expressed by such disparate materials, cultural work is enacted in interactions between human and non-human actors, artefacts, and a variety of material-semiotic arrangements. What these "arrangements" reveal is a commitment in "dialogue" to engaging audiences in a primarily instrumental understanding of scientific innovations and sweeping accounts, such as those which conflate "general societal 'progress' with technological 'advance'" (cf. Felt, Wynne et al. 2007:73).[64] Evidence of instrumental thinking is found in diverse material-semiotic objects. Presenting them to lay audiences and larger

---

[64] A pertinent example of how the exhibition promoted the sharing of instrumental knowledge that, in turn, builds trust in technoscience's promise within and across communities and society, is the "dialogue opportunities" which it created. In providing educators of youth audiences with a number of questions (two questionnaires, one consisting of 12, and the other of 24 questions) visitors were asked to specify the miners' patron saint, the access tunnel's length near the village of Amsteg, the name of the villages where the tunnel ended, or the company which built the tunnel boring machine etc. No feedback forms were given to the 60,000 visitors, groups, teachers, students, soldiers, families, senior citizens or foreign travellers who came to see the exhibition between 2008 and 2010. No surveys were conducted during this period, nor were any planned at the time of my inquiries. Personal mail dated 15 February 2011 from Maurus Huwyler, deputy media spokesman, AlpTransit Gotthard Ltd. It is certainly fair to state that the exhibition's attempt to make meaning of technoscience and society together with primarily young audiences in promoting a type of instrumental reflexivity that occasionally even voices a heroic undertone is not unproblematic.

publics raises questions with regard to existing science and governance cultures. These cultures appear to sustain dominant idioms of control, and obstruct important dimensions of learning. With regard to our "deliberate" agreements as to *what* technoscientific knowledges should be publicly illustrated and performed, which is always, as Felt, Wynne et al. argue, beset with contingencies (2007:69), the question is: How do our premises of control, social or technical, and greater instrumental power (Ibid.:70) impinge on worlds such as the Gotthard Base Tunnel exhibition? In order to answer the question I take a closer look in these conclusions at one of the objects, the "human non-human" (as I want to call it) video installation of the miners and engineers (Figure 2.6). My intention is to highlight the wider context of political as well as disciplinary discourses in which cultural work is embedded.[65] I try to argue that the testimonies of the Gotthard Base Tunnel workforce about personal matters such as complacency, pride and courage do not germinate in a societal vacuum. In other words, the displayed mentality of the tunnel workers cannot be quarantined from the realm of foundational discourses that underpin industrial society's practices of specific knowledge-cultures.[66]

As *everything* is part of a larger political and discipline-oriented framework, it can be argued that each human testimony, including the labour of the non-human actors, namely the machines, in some sense *performs* and *reproduces* technoscience. The testimonies, it is implied,

[65] Governmental, institutional and economic interests are connected to one another on the level of transnational interests and the expected impetus for economic growth. Cultural work plays a central role in catalysing and linking public interest and discourse around the civic issue of building the tunnel. Future European rail mobility is the driver of everything, as the construction of the Gotthard Base Tunnel constitutes an important element in the modernisation of the Paris–Dijon–Dole–Lausanne/Neuchâtel–Berne rail line, and the Rhine-Rhone high-speed rail route approved by Switzerland's Federal Council. It will make train travel between Switzerland and Paris 15 minutes faster. Swiss Federal Office of Transport (Ed.) 2007. Through the Alps at 280km/h. *Alptransit. The Future of Rail*, 2, March. p. 2.

[66] An example given by Margaret Wertheim is the "Theory of Everything" (TOE) in physics that, in her view, is more of a "quasi-religious (...) than a scientific goal" (1997:13).

sustain technoscience's foundational narrative, for which the Alps (nature) are the "raw material" for human action. By drawing on Bourdieu and Passeron (1973), we may even be tempted to say that the testimonies formally represent the *cultural capital* required to reach out for the economic benefits that the building of the tunnel promises. Moreover, we may now see that the entire workforce *and* the cultural workers themselves, who have constructed this techno-socio-cultural knowledge space for public reflection, are altogether subjected to governmental control and implicated in the reproduction of regimes of power in governmentality perspectives, thus playing a *crucial* role in the exercising of power. In other words, as social subjects they are "socialised into an acceptance of the virtues of rationally managed systems of work" (Banks 2007:45), and, while the Gotthard Base Tunnel exhibition denotes the technical and instrumental, its creators, as "active subjects", appear to have become what Banks calls the "object and target" of disciplinary discourses (Ibid.:46).[67]

Similarly, the aura of the little statue of the Christian patron saint mirroring her allegedly implicit faith in the doctrine of the Holy Trinity, the central dogma of Christian theology, contributes to stabilising preferred visions of a socio-cultural and religious order. It can further be said that with the decision to place the Christian patron saint of the miners front and centre, the curators of the Gotthard Base Tunnel exhibition make reference to the spiritual dimensions of Western culture by excluding—among other salient issues—actual or past references to the contentions of religion and

[67] Individuals are "somehow deeply implicated in the reproduction of regimes of power", Banks notes (2007:46). Drawing on Foucault, he writes: "[W]e are alerted to the relationships between social power and discipline of individuals, and the ways in which the political-economic framework of societies are [sic] not simply dependent upon the formal government of abstract systems, but on the management and control of individual practices and the self. This means that (as Foucault argues), primarily in the interests of capitalism, the micro-processes of social reproduction, right down to the fine detail, of for example individuals' work, health, family and sexual conduct have become the focus of discipline through calculated mechanisms of administrative control (Ibid.:45).

politics. In sum, the inquiries into a number of material-semiotic objects and their essential qualities have offered insights into their agentic potentials and powers. However, as we have seen, this is not an issue exclusively limited to the problem of how to represent, mediate, translate and perform "reality". It is more than that. As a complex cultural system of knowledge and information, the Gotthard Base Tunnel exhibition reveals from the perspective of a deeper analysis how the humanities and the social distribution of knowledge are commodified and marketised today in initially non-transparent ways (cf. Gibbons et al. 1994:91, 93, 95).

Furthermore, from within the framework of my ethnographic observations and translations of the human and non-human assemblage, the contemplation of the different material-semiotic objects offers a way of *understanding*, and a way of *knowing* about our staging of things. This relates to a further understanding of the politics of knowledge, and thus opens a horizon for other ways of relating, creating and imagining (*sensu* Maria Puig de la Bellacasa [2011:99]).[68] It is a confrontation with the way we represent things, the way we convey knowledges about these things, and it helps to understand how things are held together: how religion, for example, is materially and semiotically represented and what implications there are across the bifurcation of consciousness. And it is also an approach to what Puig de la Bellacasa refers as an *ethicality* that encompasses socio-technical assemblages (Ibid.:100).

Overall, the exhibition is conceived as the outcome of messy practices of materiality and relationality, and the result of mediating knowledge and socio-political priorities through specific semiotic apparatuses. The

---

[68] Maria Puig de la Bellacasa is a researcher with key interests in the development of feminist knowledge and practices that question traditional modern Western knowledge.

result of these arrangements could be different in terms of different relationalities, a different reflexivity and ethicality, and different pragmatics.
One of the key issues that arises from such critical reflection is, I suggest, that different stories about science, technology, epistemology and
politics are possible. While the actor-network approach contributes in
this case study to a better understanding of the very "essence" of society,
it sensitises us at the same time to the possibilities of enacting different
realities/priorities, and to going further instead of assembling "existing
concerns" and being obsessed with power, as Puig de la Bellacasa outlines (Ibid.). Finally, accepting that the Gotthard Base Tunnel exhibition
and its primarily instrumental realities constitute what I would call a
"techno-hyped space" should sharpen our attention to the question as
to how the representation of science and technology, and our staging
of things, could engage us in more reflexive ideas of knowledge and
learning.

To conclude, the descriptive analysis has served to interweave theoretical understandings of the agency of cultural work in the particular
material-semiotic setting of a civic dialogue system. It is important to
understand that Actor-Network Theory has served in this study as a distinct tool for the analysis of the material-semiotic properties of different
forms of knowledge, and the heterogeneous relations of socio-economic
and diverse invisible "networks" that reproduce regimes of power. Both
the human and the non-human actors are embedded in these "networks",
which are stabilised by the actors and which empower the actors themselves. The problem with the descriptive approaches of Actor-Network
Theory is, as Walsham points out, that these studies produce a "veritable
mass of detail", and "book-length output" (1997:476). Latour, Walsham
further points out, is aware of the theory's limitations in trying to identify all the numerous associations between human and non-human actors

(Ibid.). Yet, while Actor-Network Theory emphasises the importance of detail, for presentational motives the problem of selection, according to Walsham, "tends to be magnified" (Ibid.). As a consequence, Walsham proposes to experiment with different ways of describing case studies in "paper-length format" (Ibid.). I should mention at this point that with my more than paper-length study and the methodology of reflecting on the "reality" of material objects and their semiotics, I have tried to contribute to a specific awareness and sensibility. The ANT perspective has helped to understand "how" realities and relations are constructed and performed within a public display. However, the case study could have been based on a completely different methodology, one, for example, that would have explored the origin of the concept, design development, approval process, funding and the assessment of its final presentation etc. Instead, the actor-network approach has offered insights into material-semiotic practice, meaning and materiality, and the various kinds of human and non-human agency this entails. Furthermore, the case study has revealed "how" a knowledge space is created through cultural work, and has sensitised us towards what Law calls the "messy practices of relationality and materiality of the world" (2009a:141).

As we become more and more dependent on scientific knowledge's instrumental powers in technology, developing more reflexive dimensions of collective awareness becomes increasingly important, not less so, Felt, Wynne et al. argue (2007:64). This means, I suggest, that new ways should be found to foster forms of "reflexive" thinking. I thus see the following questions arising: What strategy should cultural work be oriented towards in creating more "reflexive" knowledge trajectories? What is needed to enrol citizens in processes of (self-)reflexivity and learning as well as collective imagination? Are we willing to deliberately

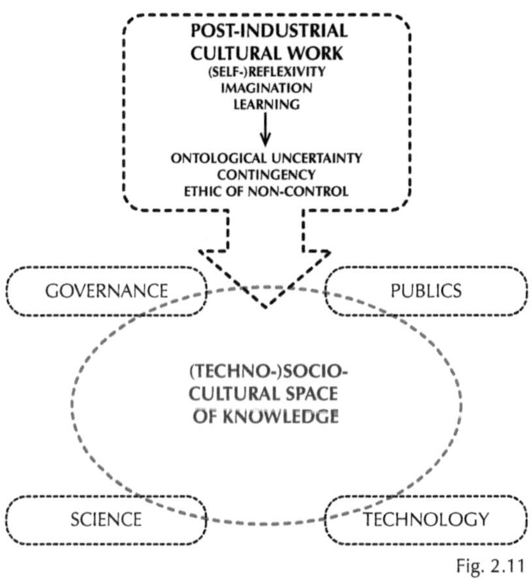


Fig. 2.11

cultivate new dimensions of learning as outlined in these conclusions? And are we willing to acknowledge diverse contingencies, ontological uncertainty as well as an "ethic of non-control" to begin discussion of the implications for governance, science, publics and technology? Figure 2.11 depicts these issues. Post-industrial cultural work could play a central role in linking public interest, and accomplishing inherently critical and reflexive tasks. Greater reflexivity could include the risks of staking too much too early on the supposed certitude and exclusivity of existing knowledge and "roads not taken", as Felt, Wynne et al. note (Ibid.:64). In addition to the need to consider new perspectives, the contingency of scientific knowledge itself could be considered for potential use in what these authors envisage as "public arenas of all sorts, whether innovation and technologies, or regulatory policies, or combinations" (68).

There are different ways of programming a public knowledge space for learning. My cut into the ontological flesh of the Gotthard Base Tunnel exhibition has laid the foundations for ongoing inquiries. In the next chapter, I will explore cultural work from the perspective of Jennifer Baichwal's documentary film about China's industrial and economic revolution. My inquiries into *Manufactured Landscapes* (2006) and Edward Burtynsky's photographic work, which documents China's industrial age in the film, reconsider the issue. At stake is the problem of how to construct and disseminate knowledge about issues of unsustainability in the age of globalisation. A propositional taxonomy for post-industrial "cultural knowledge work" (CKW) will then be sketched by drawing on the previously elaborated findings.

## *Manufactured Landscapes*—Manufacturing a Two-Dimensional Cultural Space of Knowledge

> The problem is that transparent, unmediated, undisputable facts have recently become rarer and rarer. To provide complete undisputable proof has become a rather messy, pesky, risky business. And to offer a *public* proof, big enough and certain enough to convince the whole world of the presence of a phenomenon or of a looming danger, seems now almost beyond reach—and always was. The same American administration that was content with a few blurry slides "proving" the presence of non-existing weapons in Iraq is happy to put scare quotes around the proof of much vaster, better validated, more imminent threats, such as global climate change, diminishing oil reserves, increasing inequality. (Latour 2005:9)

In this case study, Jennifer Baichwal's documentary film *Manufactured Landscapes* (2006) will be exposed to the unambiguous question whether the film and, in a broader sense, the "cultural production of knowledge" have any "social", "political" or "epistemological" significance. I will reflect on both the representation of things in the film, and the concerns of cultural workers regarding the environmental, ecological and social consequences of China's industrial revolution. At stake is the "knowledge" that *Manufactured Landscapes* offers, as well as ethical and moral issues in relation to the film, and their neglect in cultural production today (Banks 2007:102). I will discuss these issues in my conclusions in the light of the instrumental pressures to which cultural practices and "reflexive" learning are exposed today. The challenge of this case study is to assess the knowledge that *Manufactured Landscapes* offers in the context of the "politics of the present" (Ibid.:165), and to address both the concerns of the film in contributing to our awareness and the agency of cultural work. Latour's quote above, referring to Colin Powell's (the former US Secretary of State) "infamous talk" about the "unambiguous, and undisputable fact" (Latour 2005:8) of the presence of weapons of mass destruction in Iraq, provides the starting point for my inquiries.[69] Faced with Powell's failure, which debunks the demoralising face of politics as well as Latour's hope expressed by the words "Can we do better?" (Ibid.:9), I ask: Can the humanities do better instead of being what Gibbons et al. call "quizzical commentators" who offer "doom-laden prophecies", "playful critiques" and "pastiche entertainment"? (1994:110).

Jennifer Baiwal's film and its real-world conjunctions constitute a breath-taking cultural response. I will explore this narrative of the disturbing grand scale of China's industrial revolution as an assemblage of environmental, cultural, social and ethical issues—the result of cultural work. In an attempt to explore practices and their specificities, which enact realities and the provision of knowledge, the empirical material will serve to reflect on the relational entity created by the interaction between human actors and a non-human "actor"—the film. This interaction is conceived as an entangled entity—a process of stabilisation and translation of "realities" in its own right. At stake are practices of representation and *matters-of-concern* that Baichwal and Burtynsky express in seeking to reach audiences.[70]

---

[69] The talk was given at the United Nations Security Council meeting at UN headquarters in New York on 5 February, 2003.
[70] My use of the term *matters-of-concern* draws on Latour's distinction between *matters-of-concern* and *matters-of-fact* such as those presented in the political context at the United Nations Security Council regarding the supposed presence of weapons of mass destruction in Iraq.

## Methodology

I will proceed as a kind of photo-epistemologist, reflecting on the "deeper" meanings of the film. My intention is to make *Manufactured Landscapes'* intrinsic engagement more transparent. It is quite obvious that my way of proceeding methodologically, that is, how the material-semiotic representations will be discussed from within the broader framework of the present social, cultural, economic and political realities, inevitably raises questions regarding our "knowledge" of the world. *How* does it build and configure reality? *What* is its sense of reality? My analysis is not based on a physical "field trip" such as the one on which I have embarked in the Gotthard Base Tunnel exhibition. With the aim to shed light on Baichwal's and Burtynsky's two-dimensional "space" of knowledge, I have selected 29 film stills. Everything I will reflect upon is subject to Baichwal's and Burtynsky's distinctions (translations) between issues and non-issues, actors and non-actors. Initially following more descriptive and occasionally analytical approaches again, I will contemplate the film stills as if they belonged to an imaginary exhibition. While I study them as material-semiotic entities echoing the political, technological and cultural consequences of economic globalisation, I present them in a particular order that is not consistent with the actual narrative of the film. I will use excerpts from an interview given by Baichwal and Burtynsky in the American *Bright Lights Film Journal* to contemplate some of the challenges that the making of this film entailed.[71]

To begin, I follow Baichwal's camera, reflect on different scenes, and contemplate Burtynsky's comments and photographs taken during explo-

---

[71] <http://www.brightlightsfilm.com/about.html#they> downloaded 28 February 2009.

Fig. 2.12 (Film Still 16'38")

Fig. 2.13 (Film Still 57'20")

Fig. 2.14 (Film Still 13'18")

Fig. 2.15 (Film Still 16'20")

rations into what he calls "manufactured landscapes". Initially, these "landscapes" are slag heaps, e-waste dumps and gigantic factories in the provinces of Fujian and Zhe-jiang. Figures 2.12, 2.13 and 2.15 are black-and-white footage taken by Burtynsky's cameraman during previous visits to China and introduced by Baichwal into the film. Figure 2.12 shows a girl in the midst of piles of metal rubbish in a rural village. This girl is wondering about a polaroid photograph of the scene taken by Burtynsky. In Figure 2.14 menial labourers scour sky-high piles of metal junk and electronic waste for copper filaments that originate from discarded PCs and circuit boards shipped over from America. Burtynsky comments on the subtext of these photographs:

> Everything has consequence, and this is about consequence.
> This is about the other side of our built environment and the
> other side of our consumer culture. There's this other world
> that is massive and ever-growing, and it has consequences
> both to the diminishment of natural resources and to the
> expansion of China and the externalization of a lot of the
> dirty stuff that it takes to make the stuff we like. But it's going

Fig. 2.16 (Film Still 53'36")  Fig. 2.17 (Film Still 52'51")

Fig. 2.18 (Film Still 47'23")

into their rivers, and into their air, their food and water, and so all those things are on the table.[72]

Figures 2.13, 2.16 and 2.17 show urban locations in the Three Gorges Dam area, where the Chinese government realised the world's largest hydro-electric project in 2004, displacing 1.1 million people from thirteen cities, 140 towns and 1,350 villages.[73] Figure 2.18 depicts a worker on the gigantic Three Gorges Dam construction site. In Figures 2.16 and 2.17 we see residents dismantling their soon-to-be-flooded cities. In Figure 2.13 a man again shows a polaroid photograph on which he finds himself surrounded by demolished houses that once belonged to one of the thirteen Yangtze River cities.

Baichwal's open-ended approach allows viewers to think about their moral concerns prompted by these pictures and their attendant ethical dimensions. The ethics and morality, I suggest, of both the Three Gorges

[72] Smith 2007, no pagination.
[73] <http://www.internationalrivers.org/campaigns/three-gorges-dam> downloaded 8 February 2013.

Fig. 2.19 (Film Still 7'44")     Fig. 2.20 (Film Still 6'44")

Fig. 2.21 (Film Still 3'58")     Fig. 2.22 (Film Still 5'08")

Fig. 2.23 (Film Still 27'46")     Fig. 2.24 (Film Still 28'09")

Dam builders (the Chinese government), and Burtynsky and Baichwal as cultural producers of these images, become strikingly apparent. In Figures 2.19 and 2.20 the masses of lined-up workers at Cankun Factory have their pep talk in the morning. Figures 2.21 and 2.22 show the film's opening scene: a nine-minute tracking shot of a gigantic work floor. Figure 2.21 is a shot of an aerial long view of this factory floor where 20 million irons are produced per year. This is Burtynsky's and Baichwal's "theater of industry",[74] a heterogeneous amalgamation of economic, social and political ingredients offering an ambivalent awareness of the agency of humans.

[74] Smith 2007, no pagination.

Fig. 2.25 (Film Still 34′24″)

Fig. 2.26 (Film Still 31′52″)

In Figures 2.23 and 2.24 we see a Chinese shipbuilding yard. Burtynsky contemplates these elephantine ships under construction, symbolising the awe-inspiring scale of economic globalisation, material production and global consumption. A "great metaphor", we hear Burtynsky saying, as he points to both these ships that "connect us through seas", and the all-encompassing matrix of a globe-spanning economic world. For Burtynsky, these vessels *are* the consequences of our ways of reasoning allowing globalisation to take immense proportions. "Through [these] ships", he reflects, come all the materials we experience.

While Burtynsky's thoughts offer an insight into the world's networked dimensions, his images provide visual evidence of the side-effects of economic globalisation. Yet, his reflections stay somewhat aloof from what we may call the *true face* of Western industrialisation—the side-effects of Western economic power and wasteful lifestyles.[75] Deliberating on the abandoned shipwrecks at Bangladesh's Chittagong Beach[76] (Figures 2.25, 2.26, 2.27), he muses:

One time, I was photographing a silver mine. I rode in my
car made of iron and filled with gas [fuel]. I took my metal

<hr>

[75] Vandana Shiva's term for the consequences of Western industrialisation is *environmental apartheid*, in which, she writes, "through global policy (...) the Western transnational corporations supported by the governments of the economically powerful countries attempt to maintain the North's economic power and the wasteful life-styles of the rich" (2000:113).
[76] Chittagong Beach is one of Bangladesh's ship-breaking beaches.

tripod, grabbed a film that was loaded with silver [silver halides] and started to take pictures. Everything I was doing was connected to the thing I was photographing. Looking at these ships in Bangladesh the connection for me was clear. At some point, I probably filled a tank of gas from the oil that was (...) delivered by one of these tankers.

## Conclusions

The film is the result of the self-producing strategies and perseverance that Baichwal and Burtynsky have employed as cultural producers (Fig. 2.29, 2.30).[77] It features open-ended contemplations on the consequences of globalisation, and makes the troubling downside of mass production and environmental degradation visible. China's industrialisation serves as an example to illustrate the impending ecological crisis. These issues are distilled in impressive images such as a vast coal-distribution field (Fig. 2.31), the ready-to-be-shipped containers in one of China's monumental harbours (Figure 2.32), or a gigantic parking lot of newly manufactured cars (Figure 2.33). These images contrast with those of Bangladesh's poverty-stricken rural population posing for Burtynsky's camera amidst

---

[77] Baichwal comments on the problems they faced during filming: "We wanted to talk to more people, but whenever we tried, we got into trouble. Their supervisors said: 'Oh, you're bothering them, they're trying to have their lunch'—meaning: 'Don't talk to them'. We interviewed this woman at the shipbuilding yard who was a welder [Figure 2.28]. She looked like she was 18 years old and I thought: What are you doing here? Why aren't you in school? (Laughs) She said, 'I failed to get into high school so I came here to work; my mother worked here', and it was a fairly benign conversation. But we were not allowed to go back there the next day, because she was not the official spokesperson for this place and was not giving an official story. Her view was meaningless and therefore should not have been solicited. So that dance was constant. And the sequence (at the) coal-distribution (field) [Figures 2.29, 2.30, 2.31]—it's always difficult to be self-referential in film. It's a tricky balance. I kept coming (into the frame) [Baichwal is hidden behind Burtynsky on the left side in Figure 2.29] because I wanted to show the difficulties that Ed [Edward Burtynsky] goes through (negotiating with officials). I wanted to give a sense of that, but at the same time didn't want it to be 'Poor us, the film crew that's getting oppressed'". Smith 2007, no pagination.

Fig. 2.27 (Film Still 33'31")

Fig. 2.28 (Film Still 28'05")

Fig. 2.29 (Film Still 42'03")

Fig. 2.30 (Film Still 42'45")

barrels of sludge gathered by hand from rusty oil tankers at Chittagong Beach (Figure 2.34).

My analysis of *Manufactured Landscapes* as a material-semiotic gesture as well as the outcome of cultural work has conceived of the film as an entangled entity of human actors and a non-human actor—the film. From the perspective of Actor-Network Theory and its non-foundational world, *Manufactured Landscapes* can thus be seen as an assemblage of the material-semiotic engagement and "reality-work" (Law 2009b) of cultural actors, who have created a two-dimensional "space" of knowledge. In other words, the theory has provided, on the one hand, a way of looking at disregarded practices that generate the social, while, on the other hand, it has helped to understand both the significance of materiality, and the self-assignment of cultural workers in targeting publics, their concerns and imaginations. It is important to note that in this case study the ANT perspective has helped to understand the semiotics of cultural production—"how" the social "is done" and "how" people are enrolled in processes of collective thought. The number of "objects" discussed is of secondary significance. What is more important is to grasp the material-

Fig. 2.31 (Film Still 43'33")   Fig. 2.32 (Film Still 25'02")

Fig. 2.33 (Film Still 37'00")   Fig. 2.34 (Film Still 36'05")

semiotic dimensions and materiality of cultural work to enact realities for civic reflection.

While collective imaginations could play a more decisive role in the humanities and arts, *Manufactured Landscapes* appears to *subvert* consciousness as it prevents us from seeing the deeper implications of the much larger global problems that we are facing. Unlike most film documentaries there is very little commentary that would allow viewers to *make sense* of what the film's troubling images imply, such as the disconnection between collective action and the risks they involve.[78] My analysis therefore poses general yet far-reaching questions about the deeply ingrained assumptions that have come to shape our Western ideals—the "hollow promise of fulfillment and happiness through material gain", as Burtynsky notes.[79] Moreover, the review of issues pertaining to the more discounted potentials of the humanities and arts as an action field for

[78] Kenneth Baker writes: "'Manufactured Landscapes' leaves its audience with many troubling questions. Among them: Should a film console us with its own brilliance when it aims to discomfit us with its content?" <http://www.sfgate.com/movies/article/FILM-CLIPS-Also-opening-today-2580576.php> downloaded 10 February 2013.
[79] <http://www.edwardburtynsky.com/Sections/News/Burtynsky_info_Sept_2005.pdf> downloaded 25 March 2009.

Fig. 2.35 (Film Still 71'15")

Fig. 2.36 (Film Still 70'41")

Fig. 2.37 (Film Still 69'55")

Fig. 2.38 (Film Still 10'35")

pressing human affairs becomes more important. Yet, as Goleman points out, the problem lies much deeper as it is entangled in a fundamental "perceptual dilemma"—a "vast, blind spot" that we all share (2008:32).

While the crisis appears to be linked to the material hunger of Western consumerism, which has recently been exemplified by the pace of China's industrial growth, the film's open-ended intention to make us aware of the multidimensional aspects of this hunger is based on the virtues of remoralising impulses.[80] I re-emphasise that the moral-ecological concerns and sensibilities constitute an element of the larger research trajectory of my text, which includes questions such as why cultural actors do what they do, and who they are. This encompasses and relates to human aspirations and values. While Actor-Network Theory's social constructivism and its amorality in particular, as Walsham notes (1997), have been criticised together with its being indifferent to judgment and political biases concer-

---

[80] Regarding our Western ideals, Burtynsky writes: "The mass consumerism these ideals ignite and the resulting degradation of our environment intrinsic to the process of making things should be of deep concern to all. I no longer see my world as delineated by countries, with borders, or language, but as 6.5 billion humans living off a precariously balanced, finite planet." <http://www.edwardburtynsky.com/Sections/News/Burtynsky_info_Sept_2005.pdf> downloaded 25 March 2009.

ning the choices that social actors have (Ibid.:473-74), the theory has provided a way of coming to terms with both the more reflexive dimensions of collective life and the agency of cultural work. Hence, the ongoing questions are whether cultural work (and cultural workers) will be able to adopt qualities that may have a transformative effect and which new avenues of possibility these might open up. The current social conditions of cultural work, Banks notes, are "under-rewarded", "brow-beaten" and "drained of political will" (2007:184). Moreover, Banks points to the consequences of a form of "absolute power" that the capitalist "culture industry" has unleashed (Ibid.). The ramifying problems that this entails need some elaboration with regard to pressures, including pressures for speed from commercial and profit-oriented interests.

With regard to the reflexive ideas of knowledge and learning—a traditional characteristic of the humanities (Gibbons et al. 1994:102)—the predominance of instrumental pressures, according to Felt, Wynne et al. (2007:66), poses a challenge to the development of learning processes. At the same time it is also a crucial moral issue. In addition, the present situation with its (politically) instrumental and commercial dimensions seems to discredit non-instrumental values, and oppositional art is "decontextualized, commodified and thus (arguably) divested of its critical power", as Banks outlines (2007:157), based on Bell (1976). Drawing on Bourdieu (1993) Banks further notes that, today, cultural workers are "primarily (if not entirely) driven by instrumental, status-seeking (and economically acquisitive)" endeavours (2007:161), or even by an ostensibly "moral" awards system. As a consequence—and in view of the regime of the present capitalist-oriented power-knowledge relations of the humanities and arts—cultural work appears to be rather precariously positioned. One could therefore pose the question whether Baichwal and Burtynsky's contributions to the processes of reasoning have any impact on the recreation of our awareness regarding the proof of issues such as the more imminent threats of global environmental collapse, diminishing oil reserves or over-consumption etc. If we accept that the vocation of the humanities, as Armstrong suggests, should have *"deep implications* for the economy, politics and society as a whole" [my italics], and is to "get those central areas of life to go as well as possible for as many people as possible" (2009, no pagination), then *Manufactured Landscapes'* engagement is perhaps at stake. The deeper question is how to embrace knowledge with a central concern for human beings and our self-reflexive awareness of *how* we know what we know, and *how* we act.

From the perspective of Actor-Network Theory, Baichwal's and Bur-tynsky's "knowledge-claims" are situated in socio-material-semiotic practices that involve knowledge and raise questions about both the reflexivity of matters of collective concern as well as the ethicality of cultural work. By claiming that the notion of matters-of-concern is rela-tively new, but "the concerns that support it are not", Puig de la Bellacasa problematises both knowledge politics and the ways we present things by pointing to the notion's "subtle, yet meaningful displacement" (2011:87).[81] In short, Baichwal's and Burtynsky's presentation of concerns such as the demolished Yangtze River cities, the dehumanising scenarios at Cankun Factory, the exploitation of humans on Chittagong Beach, and the awe-inspiring dimensions of China's industrial revolution does not allow for "ready-made" explanations in view of my cartography of the actors and the realities conveyed. Yet, the engagement of these actors telling "stories" through constructivist accounts raises questions of how to practise what Puig de la Bellacasa describes as a *respectful ethos* of knowledge pro-

---

[81] Puig de la Bellacasa argues that, in contrast to "interest"—"[the] previously prevalent notion in the staging of forces, desires and the politics sustaining the 'fabrication' and 'stabilization' of matters of fact"—"'concern' alters the affective charge of the thinking and presentation of things with connota-tions of trouble, worry and care" (2011:87).

duction (Ibid.). Part of the problem with the ethos of cultural work is that cultural workers in reflexive modernity continue "to fasten on the culture-masks proffered to them and practise themselves the magic that is already worked upon them", as Banks notes, quoting Adorno (2007:160; Adorno 1991:82). For Banks, the question whether cultural work has "progressive or transformative potential" thus remains open (Ibid.:184). A more inclusive understanding of cultural work is perhaps obtainable, he notes, but can "only emerge through further detailed investigations into the structures of cultural production—including more thorough assessment than hitherto of the motivations, practices and actions of cultural workers themselves" (Ibid.).

In order to properly understand under which conditions the public "knowledge space" of *Manufactured Landscapes* was generated and to deliberate on such conditions in the subsequent chapters, it was necessary to engage with the social, ethical and "reflexive" dimensions of cultural work and governance. As Gibbons et al. note, there are different forms of learning from the dominant culture of instrumentalism and the ways in which we can manage to recognise reality through our knowledge are entangled in social, moral and epistemological aspirations that the humanities and arts have traditionally nurtured (1994:7). In its conclusions, Baichwal's documentary film indeed ignites moral-ecological viewpoints, yet its "dialogic" engagement appears to be at odds with the *late realisation* of compelling evidence regarding the economic and ecological dilemmas that the modern existence of humans has created (for example, the unsustainable life cycle of non-humans depicted in Figures 2.39 and 2.40). I have always respected the humanities and arts for questioning the views we harbour and the ways in which we think about the world, and for structuring and sustaining ontological demands and cultural behaviours. As I have argued earlier, both the humanities and the arts depend, however, on marginal spaces, while institutions in science and technology provide ample space for the production and distribution of knowledge. Furthermore, our "dialogic" pursuits, social practices and intellectual values are bound up in humanities that, according to Gibbons et al., are

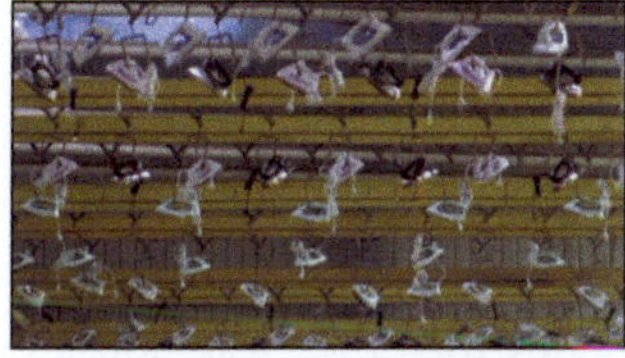 

Fig. 2.39 (Film Still 20'19")      Fig. 2.40 (Film Still 12'40")

commonly regarded as pre-industrial, even anti-industrial (1994:93). An additional problem is, as Lash outlines, that modernity and our way of organising and classifying it never corresponded to what is really going on in thought and practice, and never *recognise* the *consequences* of these practices (1999, no pagination).

In order to develop a different understanding of the problems at stake, I suggest that a *different* ethos of cultural work building on more constructively-engaged and (self-)reflexive qualities should be made available. Forms of knowledge should be seen as powerful *cultural forces*, as McCarthy emphasises (1996:10). An ongoing question is under what assumptions and conditions either of these issues might become a (political) perspective for both cultural work and the fragmented public face of the humanities and arts. The greatest challenge is that the humanities and arts are "downsized" everywhere, as O'Brien notes in the foreword of Martha Nussbaum's work *Not for Profit* (2010). Both the humanities and the arts face the "cultivation of the technical" (Ibid.:23), and there is a serious problem with their marginal standing which must be recognised as a challenge in itself. Whether the shift of emphasis from cultural work to "'reflexive' learning" will open up "deliberative questions of human ends, purposes, and priorities", as Felt, Wynne et al. suggest (2007:17), thus finally becoming a central issue, is an open question.

In the next chapter, I will sketch a preliminary taxonomy of "cultural knowledge work". In the following chapters, my aim is to further reflect on the conduct and role of cultural workers (and cultural work) as reflexive, social, moral and political stakeholders. What are the conceptual and practical challenges for cultural work to reorient the various kinds of humanities- and arts-based civic dialogues towards providing a stronger anticipating and collective awareness? Based on the complexity of the issues that have thus far been discussed, there are further important and far-reaching questions lying ahead. One particularly challenging question is what reorientation is needed for the humanities and arts that—from a post-apocalyptic point of view—might become "rekindled (...) disinterested practices of 'amusement and self-betterment'", as Banks suggests, drawing on David Harvey[82] (2007:170; 2000:273).

---

[82] David Harvey is a British social theorist whose work has contributed to the development of modern geography, political debate and the critique of global capitalism.

# A Preliminary Taxonomy of Post-Industrial "Cultural Knowledge Work" (CKW)

As outlined in the Introduction, *knowledge work* is considered to be imperative for business success in its own right, yet, given its importance, it is little understood and perhaps poorly defined. As we have seen, there exists little consensus as to its definition due to the nature and processes of work itself, which different authors have criticised (Collins 1998; Despres and Hiltrop 1995; Drucker 1991; Liu 2004). Despite these imponderables, I propose to rethink cultural work as a form of "knowledge work" since the very term can serve to connote important issues such as diverse socio-epistemic properties, collective reflection and new practices that would change ordinary ways of thinking (cf. Felt, Wynne et al. 2007). Further, while knowledge has become a marketable commodity in the economy, it has generated a great deal of interest in its significance, impact and implications.

As a point of departure, there is an etymological problem that should be addressed as the term "cultural knowledge" itself is entangled in diverse meanings and interpretations. Choo et al., for example, have suggested that *cultural knowledge* should be recognised as an "analytical category". They link the term to the domains of shared beliefs, norms, values and cognitive structures that are "habitually used (...) to perceive, explain, evaluate, and construct reality" (2000:43). Opie sees the potential of cultural knowledge in a conception which is "able to integrate the subjective or tacit, the social interactive, values, and local geographic dimensions of knowledge" (2001:6).[83] And in the US

---

[83] Opie points to the foundational significance of cultural knowledge in New Zealand alongside its recognition as an economic and social resource in the formation of the country's knowledge society (2001:1). He uses the term in a broad sense and sees it as linked to global information networks. According to Opie, cultural knowledge may be a result of the global information flows that "intersect with an acculturated location", and thus, when combined with local resources, cultural knowledge can provide, he claims, "unique possibilities for the creation of meaning" (Ibid.:9).

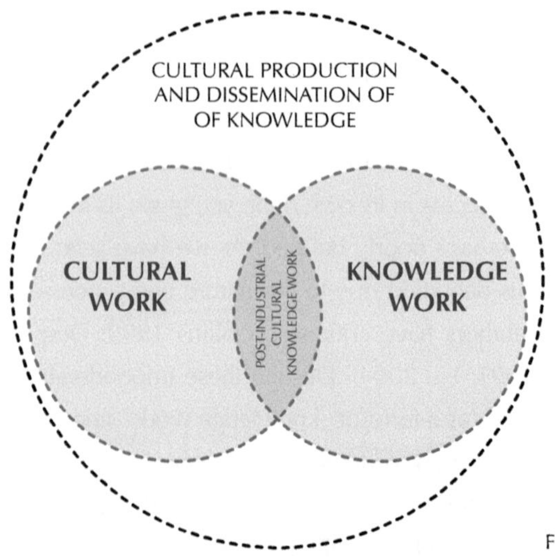


Fig. 2.41

army, cultural knowledge has been identified as a knowledge asset serving to wage successful counterinsurgencies (Miyoshi Jager 2007:10).[84] Contrary to these arbitrary understandings of "cultural knowledge" and its proximity to normative and reality-driven schemes and implications, the idea of post-industrial cultural knowledge work (Fig. 2.41) is based on a different conception. The key words are *cultural* and *knowledge work*. Put simply, the underlying core idea is to emphasise the link between cultural work and the more (self-)reflexive and (socio-)epistemic dimensions of knowledge, language and learning. The human dimensions of reflexivity, language, learning and knowledge, promoted together with diverse civic "knowledge-abilities" (cf. Felt, Wynne et al. 2007:10), should play a role in the post-industrial cultural context in which the production and dissemi-

---

[84] Cultural knowledge is primarily linked here to "knowledge and information" that plays a crucial role in strategic warfare scenarios, i.e. information about the "'human terrain'—the social ethnographic, cultural and economic, and political elements of the people with whom the force is operating" (2007:12)

nation of knowledge are becoming increasingly important. I thus relocate Banks' theoretical understanding of the *politics* of cultural work—how it is "constructed, managed and performed" (2007:3)—in a discussion of the role and impact of knowledge in domains where social, cultural and scientific learning have been essentially neglected. Yet, the open question is: What sort of knowledge should cultural work generate and convey? And is there a perspective for (post-industrial) cultural work as a distinct form of knowledge work to sustain "dialogue" processes and to cultivate what Neil Postman (1990) has called "the symbolic ecology of cultures"? In seeking to provide a rationale for an "epistemology" of cultural work, I will include in the following chapters different theoretical approaches that are central to the debate on the future of cultural work. Among them are Banks' (2007) discussion of the prospects of post-capitalist cultural work, Liu's (2004) analysis of the possibilities of knowledge work in the humanities and Turnbull's (2000) postulation of a "third" space for knowledge.

## Contradictions and Challenges

According to Banks, cultural workers are "very much at the centre of the cultural industry labour process" (2007:7). Moreover, as Banks further elaborates, "it is they who are primarily responsible for the production of those symbolic commodities judged to be essential components of the transition to a 'post-industrial', 'creative' or 'knowledge' based economy" (Ibid.:7). Cultural work may thus be seen as a "corollary of mind", yet, while the cultural production of knowledge occupies a more ambivalent and lateral space, the marginal standing of the humanities and arts themselves is reinforced. In addition to the fragmentation of the humanities and arts, this space is ruled by the motors of global economic life and many observers also emphasise the "totalizing" hegemony of economic thought and capitalist social relations (157).[85] It is therefore an open question whether cultural work in the age of corporate knowledge work and the ruling rationalities of market culture will be able to obtain a foothold in a knowledge space different from the one in which it finds itself embedded in the present activities of the capitalist cultural industries.[86] The central aim of my text—to outline the importance and practicability of a different form of collective sensitivity and reflexivity for cultural work—is thus challenged by the notion that cultural workers are perhaps the "servile and alienated victims of global capital", as Banks writes (185).

So far, much of my undertaking to examine the details of (capitalist) cultural production should be seen as the first step to understanding the complex empirical dimensions of culturally and socially distributed knowledge. My aim has been to understand how the humanities accomplish contemplative and reflexive tasks when they try to come to terms with political, social and economic reality and the reality represented by technoscientific knowledge in particular. While it is a key issue for science and technology to transform industrial society and while science and technology have apparently been successful in doing so, many

---

[85] According to Capra, our lives are significantly governed by the new networked global society and profit-oriented information flows (2002:135-57). Drawing on Castells, Capra further outlines that on an existential human level, the most worrying sign of the new global economy and its markets is that it is "a network of machines programmed according to a single value" (Ibid.:141), namely the making of profit. Castells' term *Automaton* describes this most effectively, claiming that at the core of our economies this Automaton "decisively conditions our lives" (2000:56). Thus, the global network society not only provides a potent skeleton for the repatriation of profit on a global scale, but also sustains and controls both the production and dissemination of knowledge (Gibbons et al. 1994; Willke 1998; Maasen 1999; Bender 2001) in the post-industrial "knowledge society" sketched out by Bell (1973; 1987).
[86] These cultural industries manifest themselves in the production of multimillion-dollar films or art fairs whose sales total one billion euros, making their ethos blatantly commercial. These costly ventures depend on privately funded consumption and individual entrepreneurial choices exercised through mass markets. Another example of how the cultural industries are deeply embroiled in markets is the global museum construction boom with 1,200 new museums in China alone (Pollack 2008a; Ellis 2008, both no pagination).

contributions made by the humanities to reflect upon them remain "full of contradictions" (Gibbons et al. 1994:92). The challenge that deserves attention is how to accommodate the many ambivalent and often doubtful voices of the humanities and arts. The very real challenges resulting from their apparently contradictory commitments and ways in which they currently express themselves are:

- The humanities and arts should become better at dealing with language, narrative and reflexivity.
- They should be more concerned with outcomes and anticipate conditions in the real wold.
- They should become better at engaging beyond their own defined and lateral territories.

In a still rudimentary sense, this section has considered the disparity between traditional "cultural work" and the potentials for a new form of work that I propositionally call cultural knowledge work (CKW). Throughout, the emphasis has been on a particular kind of *labourer*, namely the cultural worker whose overall acceptance, role and autonomy are particularly challenged. In *The Human Condition* (1958; 1998), her most influential work, Hannah Arendt explores the distinctions between labour, work and action by developing these categories, which attempt to bridge the gap between ontological and sociological structures. Arendt aims to uncover the possibilities and potentialities of a *vita activa* by defining the three activities *labour*, *work* and *action*, and describes four possible realms: the political, the social, the public and the private. According to Arendt, *labour* is the first of the three fundamental forms of activity that constitute the *vita activa*, which comprises all activity necessary to sustain life (for example, reproduction, obtaining food, water, shelter etc.), while *work* is the second activity. *Working* is an activity with a beginning and end, and work leaves behind an enduring artefact such as a tool, a table or a building. The third activity—*action*—takes place in the public realm. While Arendt arranges an ascending hierarchy of importance for these human activities, she argues for a tripartite division between labour, work and action, and identifies the overturning of this hierarchy as central to the eclipse of political freedom and responsibility, which has come to characterise the modern age. I will address these issues in more detail in the section «Cultural Work and Biopolitical Production». The question at stake is whether cultural work will be able to perform and cultivate new forms of freedom, and to what extent cultural workers can resist the principle of economic accumulation and capitalist production. I agree with Banks that the cultural dimensions of cultural work remain "distinct (if continually hard to define)" and cannot be understood in isolation from the social context that "significantly (if not absolutely) defines it" (2007:185).

In this section, I have also attempted to draw some conclusions based on the two case studies that have provoked questions about the cultural production and dissemination of knowledge. In order to find answers to these questions, I have followed a specific path, both empirically and theoretically. While knowledge is linked to many idiosyncratic practices, the general question is how to capture the value of the social dissemination of knowledge, and how cultural work can serve to create more intense, open and democratic dialogue. The issue is therefore whether cultural work will be permitted to (re)shape and circulate our knowledge in what I propositionally call *(techno-)socio-cultural spaces* in order to break out from an unsustainable path and the contemporary problems produced by human fallibility. The function of these public spaces in which we stage semiotics and rhetoric is that of being a cultural custodian or aggregate to materially shape our imaginations and express ourselves. My aim is to further explore these issues including the kinds of ontological and epistemological questions this entails. In the next chapter, I will explore Ai Weiwei's social "mega-performance" *Fairytale* at Documenta 12 in 2007 by shedding light on the conceptual strategy and aims of this project.

## *Fairytale*—Performing a Cultural Space of Knowledge at Documenta 12, 2007

> First inevitably comes the idea, the phantasy, the fairy tale. Then scientific calculation. Ultimately, fulfillment crowns the dream. (Ai Weiwei quoting the Russian space theorist Konstantin Tsiolkovsky [1857–1935] in the epilogue of the original project proposal entitled *Fairytales* [2007a:9])[87]

This case study focuses on how the cultural, economic, epistemological and global dimensions of contemporary art, or rather contemporary *performance art* intersect to construct what we can call a social and cross-cultural space of knowledge. In 2007, the Chinese artist Ai Weiwei[88] recruited 1,001 Chinese via his blog for the project *Fairytale* at Documenta 12 in Kassel, Germany.[89] The project was based on the idea of bringing the 1,001 Chinese to Kassel and let them act as participants in a real time performance according to the artist's concept. The overall idea was born in Switzerland when Swiss art collector Uli Sigg and Ai Weiwei met while

---

[87] In the first project proposal, Ai Weiwei had anticipated the participation of 1,001 Chinese belonging to the resettled population of over one million people from the Three Gorges area. Ai Weiwei's concept was to invite these people to Germany in order to turn Kassel into a "temporary home for over 1,001 Chinese citizens" (2007a:4). In the pre-proposal, the Chinese were expected to return to China once their destroyed cities and villages had been rebuilt by the Hubei province government (Ibid.). Chinese artist Liu Xiadong turned the resettlement into the topic of a series of paintings for which Ai Weiwei wrote an introductory text that highlights the issue and provides critical reflection on the problematic social as well as economic consequences which the resettlement entailed (2004:4-5).

[88] Chinese given names generally consist of one or two characters and are written after the family name. In order to avoid confusion, I will mostly use both the family name and first name of Chinese authors, hence Ai Weiwei. In 2011 Ai Weiwei was arrested by the Chinese authorities and accused of tax evasion and other crimes. According to Uli Sigg, the Chinese authorities have not yet given him back his passport, thus preventing him from travelling abroad. Personal mail dated 29 April 2013.

[89] According to Karin Seiz of the Urs Meile Gallery in Lucerne, Switzerland, Ai Weiwei used the blog to announce his plan. However, instead of recruiting people from the resettled Three Gorges population as previously planned, he decided to reach out for people with different professional and economic backgrounds. Seiz mentions the new challenges faced by the Chinese government during that time regarding both internet control and censorship since the world wide web had become a major social and commercial communication platform in China. Personal mail to author dated 1 March 2010.

Fig. 2.42

hiking in the Alps.[90] The project was finally realised and supported by a group of cultural and economic actors from Switzerland, who helped to fund what we may call in retrospect a socio-epistemic and socio-political experiment in the domains of contemporary art.

In my inquiries into *Fairytale,* I will primarily try to understand the social, cultural and public space that the project constructed and performed. For my analysis, I will draw on texts such as previews and reviews, media responses, interviews or cultural comments, which accompanied *Fairytale* before, during and after its realisation. The project will be explored as a specific configured moment or expression of the appara-

---

[90] <http://www.focus.de/kultur/kunst/documenta-12_aid_57109.html> downloaded 3 March 2010.

tus of modernity—a projection and cultural image that could serve as a parable of our time. I will focus on a number of issues such as the principal strategy, aims and motivations behind *Fairytale*. Drawing on matters such as the project's financial, political and ethical scope, aesthetic methodology and concept-driven objectives, I will try to answer questions concerning Actor-Network Theory's capacity to explore this assemblage of human and non-human actors. Furthermore, I will answer the question as to what extent the actor-network approach has contributed to an understanding of the practices of different actors and cultural work in particular. My interest will also be focused on the question whether *Fairytale* has raised political awareness and whether the project created social conditions for learning (cf. Felt, Wynne et al. 2007:63), as Ai Weiwei claimed. Finally, the case study attempts to address how the project has contributed to the distribution of knowledge and to what extent the power of the market impacts both cultural work and the creation of awareness in a reflexive and ethical sense.

According to Nataline Colonnello, *Fairytale* was one of the "most ambitious projects ever presented in the history of Documenta" (2007, no pagination). The project was realised during the first five weeks of Documenta 12, which opened its doors on 16 June 2007. Quoting Ai Weiwei, Catrin Seefranz describes the project:

> To bring 1001 Chinese citizens to Kassel is about creating conditions in such a way that each participant may be confronted with the ordinary experiences of others in the context of one of the most important contemporary art events. Thus, it is about personal experience, awareness, consciousness as well as direct contact with others, which includes a new awareness created in this

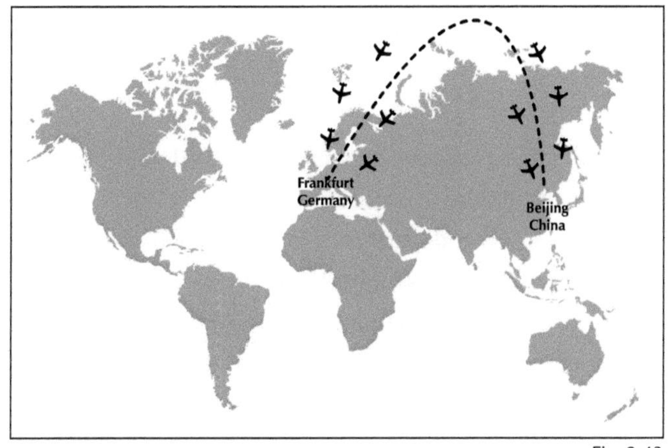


Fig. 2.43

process (...). (Ai Weiwei; Seefranz 2007a, no pagination; author's translation)[91]

The project was designed by Ai Weiwei as three interlocked yet individual projects, which, according to Charles Merewether, "extend the critical engagement with the concept of China not only in its conception of China as a physical construct, but as a constructed identity" (2007:178). The first project was the 1,001 Chinese, who were invited to Germany for one week in five groups of 200 people each (Figure

[91] Seefranz further notes: "The first call for applications on Ai Weiwei's weblog resulted in 3,000 registrations (...). The FAKE team—Ai Weiwei's temporary travel agency—will organise and coordinate the trip: participants will be reimbursed only for travel and accommodation expenses. Although in China travel to Europe is very much a privilege enjoyed by the upper classes (to work in Europe, in contrast, is not, as is quite clear from the Chinatowns and other working communities in many places), the *Fairytale* travel group will be heterogeneous and include farmers, teachers, students, artists or engineers of both sexes. The Chinese visitors will be visible as tourists: dress, town guide, maps, lunch bags, all accessories are to be designed by Ai Weiwei and his team. Whether they imitate the classic tourist uniform (basically unchanged from the times of Tati's mesmerising Monsieur Hulot with a pouch for documents) or quote those worn in the Cultural Revolution—we are not telling. One thing is certain: 1,001 guests will discover the city according to plan for around one month and will enjoy the help of guides and interpreters on their way. Another sure thing is that those travelling will document their experiences for Ai Weiwei's artistic work *Fairytale*, just as Ai Weiwei and his team will record the history of *Fairytale* in precise detail." (2007a, no pagination)

2.43 depicts the five flights, which took the Chinese to Germany and back to China).

The second project was an exhibition of 1,001 late Ming and Qing dynasty chairs distributed in clusters at different locations of Documenta 12 (Figure 2.44), and the third project was *Template* (Figures 2.45, 2.46)—a temporary wooden sculpture and architectural space for which Ai Weiwei used late Ming and Qing dynasty windows and doors, which originally belonged to traditional Chinese houses in Shanxi province, northern China, where entire old towns were destroyed. According to Merewether, the *Fairytale* project appeared to have responded directly to the three leitmotifs of Documenta 12: "Is modernity our Antiquity?", "What is bare life?", and "What is to be done?". In my inquiries, I will primarily explore the first project—the 1,001 *Fairytale* participants who came to Kassel.

## The Financial Dimensions

*Fairytale* was financed by the Swiss Urs Meile Art Gallery and two Swiss foundations, the Erlenmeyer Foundation[92] and the Leister Foundation[93] (Figure 2.47). These foundations—both private charitable bodies according to Swiss law—and the Urs Meile Gallery contributed a total of 3.1 million euros to the conceptualisation and realisation of *Fairytale*.[94]

---

[92] The *Erlenmeyer Stiftung* is now primarily involved in animal welfare. Originally established in 1981, the foundation aimed to conserve a collection of antiquities. Marie-Louise Erlenmeyer passed away in 1997. <http://www.nytimes.com/1990/07/06/arts/auctions.html?pagewanted=all>, <http://www.easymonitoring.ch/handelsregister/erlenmeyer_stiftung_271094.aspx>, and <http://www.selezione.ch/tierschutz.htm> downloaded 1 March 2010.

[93] The Leister Foundation is financed by the Leister AG, a Swiss-based manufacturer of welding equipment for plastics processing. <http://leister.com/en/leister-switzerland.html> downloaded 1 March 2010.

[94] Seefranz notes: "The dimensions of this project are 'fairytale-like' in many aspects—also in financial terms. As one can imagine, the production and realization of the project are very complicated.

Fig. 2.44          Fig. 2.45          Fig. 2.46

The funding raises two principal questions. Why did the actors behind *Fairytale* put such a huge amount into one single project? Were there any planned outcomes or any intended impact that would legitimise such an amount? Richard Dyer notes:

> The funding was generated almost immediately, and considering Weiwei's [sic] reputation in the artworld and the weight of Documenta perhaps this is not surprising. However, even though China's economy is experiencing a considerable boom, this is not an economic upturn that enriches the vast rural communities still trapped in the most abject poverty. (2007:779)

According to Dyer, the project could have taken a "very different" turn, one that would have equally involved "ordinary" citizens (Ibid.). In criticising the project, Dyer further notes: "Perhaps the notion of 'Aid as Art' might be one the artist might consider for his next project, taking into account his evident talent at raising huge amounts of funding in a very short time" (Ibid.).

The 3.1 million euros required were raised thanks to the initiative of Ai Weiwei's gallery owner Urs Meile, and through two Swiss foundations, the Leister Foundation, Switzerland, and the Erlenmeyer Foundation, Switzerland (...)" (2007a, no pagination).

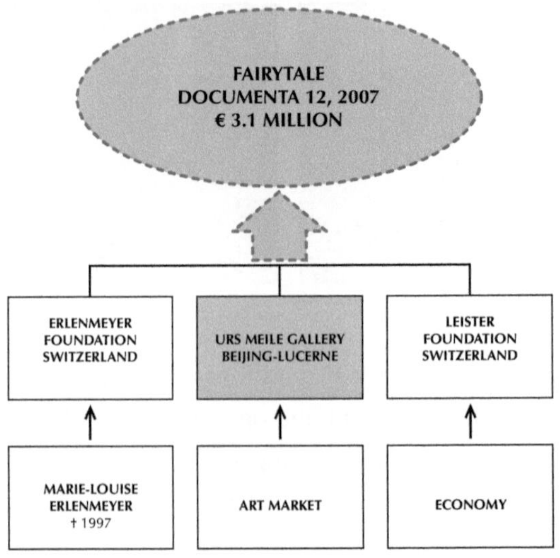


Fig. 2.47

A central question is whether the project, which for many was a "mega-performance" (Münter 2007, no pagination), a "conceptual extravaganza" (Dyer 2007:778) or just a "disruptive intervention" (Ai Weiwei 2007b, no pagination), was the result of capitalist impulses and key incentives to repatriate profit and pursue economic accumulation for Ai Weiwei. Another related question is to what extent the now global dimensions of cultural production and consumption, the power of the market, and the authority of big exhibitions impact the situation of the art world (and cultural work) with negative repercussions on the role of art in society, critical thinking as well as moral values.

## The Impact of the Global Market Culture

Until 2008, Chinese contemporary art, which had developed at a feverish pace in previous years, was the single fastest-growing segment of the international art market. Barbara Pollack provides an analysis of this development: "China is emerging as a major art center, having become a hub for buyers from South Korea, Taiwan, Singapore, Indonesia and Southeast Asia, and for overseas Chinese from all over the world. Reflecting this diversity is the wide range of foreign dealers among the 300 galleries in Beijing, including Continua from Italy, Urs Meile from Switzerland, Arario and PKM from South Korea, Beijing Tokyo Art Projects from Japan and Tang from Indonesia" (2008b, no pagination). Thus, this increasing presence of international art galleries in Beijing alongside initiatives such as the *SH Contemporary* art fair in Shanghai launched in 2007, contributed to China's new role as a global art market centre. Pollack further notes that since 2004 prices for works by Chinese contemporary artists have increased by "2,000 percent or more" with paintings that once sold for "under 50,000 US dollars now bringing sums above 1 million dollars" (Ibid.). According to Melanie Gerlis, by the end of 2006, China had already become the "fourth largest global art market by value", with a 5% share, while the US, UK and France had shares of 46%, 27% and 6% respectively (2008b:39). With a new gallery in Beijing and its massive gallery districts, 1,600 auction houses and the first generation of Chinese contemporary art collectors, the Urs Meile Gallery placed itself at the heart of these developments.[95]

As a point of departure, I refer to Yang Jiechang's[96] and Martina Köppel-Yang's[97] criticism of both the impact of the global market system's power structures on Documenta as well as Ai Weiwei's aesthetic methodology. In my conclusions to the case study, I will try to answer the more general question regarding to what extent cultural work and knowledge

---

[95] <http://www.galerieursmeile.com/about-us/gallery-statement.html> downloaded 28 April 2013.
[96] Jang Jiechang is a Chinese artist.
[97] Martina Köppel-Yang is a German art historian and curator.

production—the representation of matters-of-care as an "aesthetic and political move" (cf. Puig de la Bellacasa 2011:94)—are challenged today with regard to creating awareness in an intellectual and reflexive sense.

In a review of Documenta 12 entitled *René Block's Waterloo: Some Impressions of Documenta 12, Kassel* and published in *Yishu*,[98] Köppel-Yang first problematises both Documenta's selection process of art works and its relationship with the art market. She notes:

> On the one hand, we can see that the curators use what is convenient to obtain. Many of the Chinese artists' works in the show come from private collections or commercial galleries like, for example, Uli Sigg, Urs Meile Gallery, Vitamin Creative Space, and Gallery Loft. Moreover, each of the nine participating Chinese artists works either with Galerie Urs Meile or Vitamin Creative Space. (...) In a high-profile exhibition like this, which is considered the platform of contemporary art, works should be selected differently. (Yang Jiechang, Köppel-Yang 2007:96)

The second issue Köppel-Yang addresses is the resulting "visibility" problem of the artists. She comments on the problem with *Fairytale*'s immense budgets, which helped to increase Ai Weiwei's reputation as an already "Warholian celebrity" (Cotter 2007:24), and at the same time equipped him with a mixture of privilege and advantage to significantly advertise the *Fairytale* project in the media. According to Köppel-Yang:

---

[98] *Yishu* is a leading international art journal, which covers Chinese contemporary art and culture.

> One result of this [selection process] is an imbalance in what concerns the visibility of the artists. Again, I want to mention Ai Weiwei's example. One reason he became so prominent is that he could work *in situ* and could realise his work with a large budget. This is an enormous difference from those artists who are only represented through works, or work fragments, chosen from a collection. (...) Ai Weiwei had his own press team and thus directly addressed the media. Therefore, the media reported widely on his work and he became the real star. Most of the other artists did not work that closely with the media; therefore, their work stayed rather unremarked upon. (Ibid.)

These issues seem to resonate with Ai Weiwei's strategy that Köppel-Yang and Yang Jiechang criticise as "rather undemocratic" and basically similar to a "kind of feudalist revival". One of the problems with Ai Weiwei's methodology is the "sphere of influence" within and outside China that the artist has built up as a "local tyrant" supported by his "feudalist attitude" with negative consequences for "contemporary discourse and consciousness", as Köppel-Yang and Yang Jiechang further argue (99). According to Köppel-Yang, the particular problem with Ai Weiwei's work is, however, its "deconstructive [character] from an aesthetic point of view" (95). She emphasises that his works promoted "a kind of cliché" about Chinese culture and were perhaps driven by a "very scary" ideology with negative repercussions on the role of the individual artist and his "disappearance (...) in the process of globalization" (Ibid.)[99] Ai Weiwei's

---

[99] Drawing on observations from the 2007 Venice Biennale, the sculpture exhibition of Münster and Documenta 12, Amine Haase, editor of *Kunstforum International,* confirms Köppel-Yang's notion that—seen from a more global point of view—individual artistic agency has recently become "more and more invisible" (2007:95).

works at Documenta 12 were thus highly questionable and problematic for Köppel-Yang and Yang Jiechang. Yet, from the perspective of cultural work and *Fairytale*'s knowledge-building intentions, the deeper fundamental question is whether the cultural energy the project has unleashed has had any impact on the creation of new forms of cultural expression or learning alongside and beyond the dominant global market culture.

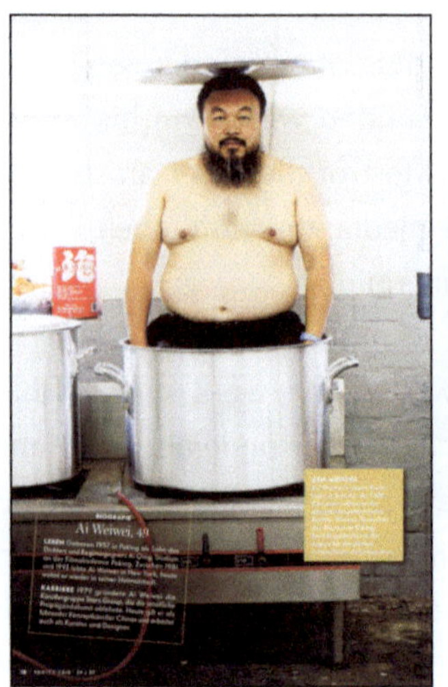

Fig. 2.48

Fig. 2.49

## An Ambitious Claim

Ai Weiwei's original concept was based on the idea of contributing to an individual, collective and global awareness that emphasised the idea of a "common horizon beyond all differences" (2007a:5). The artist's emphasis on the foundations of the individual and the collective is, however, an extremely ambitious claim, despite *Fairytale*'s intention to involve participants and the public in what Ai Weiwei proclaimed to be an experience of social and cultural innovation. Direct social involvement should thus turn visitors to Documenta 12 into thinkers about an "ethical life through art" (Cotter 2007:24). Yet, in light of Köppel-Yang's notion of the loss of the individual in the globalised market culture, *Fairytale*'s presenting the 1,001 late Ming and Qing dynasty chairs, *Template,* and the 1,001 Chinese strolling the streets of Kassel can be seen as a deconstructive move (cf. Yang Jiechang, Köppel-Yang 2007:95). Furthermore, if we accept that *Fairytale* constituted what we may call an idiosyncratic space that was based on an ideological scheme, it raises questions regarding the project's social, political and above all ethical contributions.

While, according to Philip Tinari, people in China have "acerbically" taken to calling *Fairytale* "Yellow Peril" (2007:453), for Ai Weiwei the dislocation of 1,001 Chinese to Germany was something that had less to do with art and more with an operation and construction of a different social reality as well as a different communicative structure. Hence, in David Coggin's view, Ai Weiwei's search for more "authentic" and "alternative" experiences perhaps encompasses both "clear-eyed functionality and conceptualist high jinks" (2007:125):

> To encounter the project was also to contend with the indelible factualness of it—the piece was not a proposition, these people were *here*. The participants, who applied for the trip through Ai's blog, varied widely in their ages and personal histories (ranging from peasants to poets to students), their previous travel and their susceptibility to the romance of what, for many of them, would be their first trip outside China. The piece was at once invisible—the public could not enter the living area—and extremely physical. (...). The Chinese travellers could be seen bicycling around town, playing soccer games, singing karaoke, or simply taking in the various exhibitions. (Ibid.:120)

Fig. 2.50

Thus, for Coggins *Fairytale* was—ideally—a fairy tale for the participants while the dislocation of the 1,001 Chinese mirrored Ai Weiwei's own departure from China in 1981 (Ibid.). In addition, for Ai Weiwei, the project was something far beyond what may be considered to be paid vacation for 1,001 Chinese. According to Ai Weiwei, we live in a globalised world characterised by a "new freedom of thinking", and the human actors in his cross-cultural programme became themselves a work of art that was embedded in a global and networked whole (Ai Weiwei, Jocks 2007:442). Moreover, according to Ai Weiwei, *Fairytale* was based on the intention to "stimulate" and document "experiences and knowledge gained in direct contact" between humans. The core idea was thus to filter out a narrative around new possibilities for human interaction (Ai Weiwei 2007a:3). Yet, the "humane conceptualism" and "compassionate provocation" that Ai Weiwei brought to the global scene (Coggins 2007:118) raise questions with regard to *Fairytale*'s capacities to socially and culturally shape knowledge trajectories and to generate social conditions for learning as was claimed. The resulting question thus is whether the project was at all able to catalyse new salient insights among the Chinese participants. Manuela Ammer outlines *Fairytale*'s original intent:

In thinking together the Chinese visitors and the chairs [the late Ming and Qing dynasty chairs], they generate an additional peculiar dimension alongside traditional Chinese storytelling. "Fairytale" turns into a quasi-contemporary fairy tale (...) and experimental process as it constitutes an intermediary of diverse forms of knowledge. The *Fairytale* project, which encompasses a journey, personal experiences and dialogic exchange, creates a heterogeneous knowledge environment that is different from books or images. In some sense, for the participants the fairy tale is to experience an unfamiliar world (Europe, Germany, Kassel, the international art scene), while on the personal level this leads to a different awareness and new perceptions.[100]

## Conclusions

My analysis of *Fairytale* has relied on diverse sources such as media responses, analyses, interviews or photographs. By reflecting on what other human actors have written or how they have commented on *Fairytale,* I have then based my analysis on their remarks. Some general and conceptual observations from the perspective of Actor-Network Theory are outlined below in order to clarify the practices of the different human and non-human actors behind *Fairytale* and the agency of cultural work in particular.

Actor-Network Theory's material-semiotic sensibility provides a way of making "networks" transparent in order to see how realities are enacted and

---

[100] Personal mail dated 11 July 2007 from Manuela Ammer, Curatorial Office, Documenta 12, Kassel, Germany. Translated by author.

to understand the *ontological*. Furthermore, the theory gives precedence to addressing the "making of knowledge"—the *epistemological* (Law 2009a:154). The actor-network approach thus serves to explore *Fairytale* as an experiment about knowledge acquisition, "dialogic" exchange and what we may call experiential learning. It allows for broader explanatory approaches to understanding this assemblage of ontological, epistemological and ethical issues in relation to the *knowledge space* that the project created.[101] This space was realised by bringing the 1,001 human actors

Fig. 2.51      Fig. 2.52      Fig. 2.53

from China to Kassel. As a consequence, we can say that *Fairytale* was, on the one hand, enacted through socio-material-semiotic practices such as those announcing the project on Ai Weiwei's blog, which recruited the 1,001 participants, funded the entire project, organised public transport (airplanes, buses etc.) and infrastructure (accommodation, food etc.), shipped the 1,001 late Ming and Qing dynasty chairs from China to Germany, and realised the wooden sculpture *Template*. On the other hand, *Fairytale* can be seen as an assemblage materialised by these different actors. From the latter viewpoint, *Fairytale* can be considered as a socially and materially complex engagement as well as the "collective effort" of

---

[101] As outlined in the Introduction, the sociological notion of actors as "social entities" is not entirely given up in this work, and what actors ("cultural workers") do is therefore essential for an understanding of the constructive knowledge roles that human actors play (cf. Felt, Wynne et al. 2007:75).

human and non-human actors based on particular intentions and interests. With regard to the agency of these actors and all the things involved, it is not possible to separate Ai Weiwei, the curatorial team and technical staff of Documenta 12, the employees of Urs Meile Gallery or the actors behind the two foundations from the many different practices that shaped *Fairytale*'s socio-material arrangements and actions (cf. Law 2009b). While Ai Weiwei's works and the 1,001 Chinese played a central role themselves, the human actors constituted the key element of the entire

Fig. 2.54                                                                Fig. 2.55

project alongside the socio-epistemic-political space that *Fairytale* generated and in which these actors performed.


While Documenta left many visitors "lost and confused" (Searle 2007:23), the *Fairytale* project raises questions with regard to its social and political contributions as outlined earlier. From the perspective of ANT, *Fairytale* constructed a social and epistemological space in order to institute the individual and the common. Yet, it can also be seen as being primarily a matter of self-serving endeavours by cultural and economic actors. This is an issue with wider ramifications including the impetus for writing this book, namely to understand deeply embedded knowledge practices and preoccupations that shift arbitrarily under contingent conditions (cf. Felt, Wynne et al. 2007:70). In other words, the politics of knowledge is an important element to which cultural work (cultural workers) can and should be linked in order to (re-)create awareness in a reflexive and ethical sense. The central challenge is what Puig de la Bellacasa describes as the "caring [for] relationalities in an assemblage" (2011:94). Her emphasis on the political and aesthetic dimensions of caring for knowledge production by representing matters-of-care (Ibid.) raises questions with regard to which forms of agency can be deployed, which new knowledge trajectories should be developed in public arenas of all sorts, and which issues or questions should be addressed (cf. Felt, Wynne et al. 2007:57, 67).

While a "knowledge space" can take any shape a curator or an artist wants to give to it, *Fairytale* created—somewhat paradoxically—social conditions for the participants and the population of Kassel to meet the "unknown" (Colonnello 2007, no pagination). What is at stake is not so much the creation of an experimental space, which may be seen as an opportunity for human exchange to allow open-ended potentials and capacities to evolve. However, the deliberate engagement of 1,001 human actors in a socio-political experiment with the ambitious aim of galvanising cognitive potentials as a seemingly innovative method of prompting self-reflexive behaviours should be seen as problematic. Moreover, if we reflect critically on *Fairytale*'s ethical concerns, the project's strangely ambiguous intention to sophisticate the individual as the "bearer of moral responsibility for the betterment of society" (Birnie Danzker 2008:19) seems to have contributed little to the sort of potentially enlightening and reflexive "social learning" dimensions required to overcome present inertia. In contrast, Ai Weiwei seems to have been convinced that, with *Fairytale,* he *has* contributed to a new ethics of freedom, individual thinking and more authentic forms of life when he comments that

[life] is based on individual reasoning, an individual style, individual approaches and methodologies. In my opinion, the individual system and the kinds of individual lives people live is what has remained and is of great significance to express individuality. Humans who are not able to experience an individual life are like corpses. The task of life is to be authentic with our individual lives. (Ai Weiwei, Jocks 2008:239; author's translation)

To some extent, *Fairytale*'s socio-material-semiotic engagement may have provided conditions that support the development of individual and collective sensitivities towards learning and knowledge among the 1,001 Chinese participants and the publics involved. However, the project appears to have been exposed to the larger processes of the commodification and gentrification of public space (cf. Banks 2006:463). This includes the impact of capitalist social relations, which affect the moral foundations of human engagement. Precisely these issues are what Yang Jiechang and Köppel-Yang address when they criticise the negative effects of the global dimensions of cultural production and the power of the market on the role of art in society. According to Köppel-Yang:

I feel it is the sad result of a larger development within the globalised art world where the power of the market becomes more and more important. (...) Actually, the situation of the art world does not differ from that of the field of politics and the economy. With globalization, the influence of politicians and leaders from democratic countries becomes less and less important, and that of financial interests, on the contrary [sic], increases. In the art world there is a similar development (...): the loss of spirituality, of a critical attitude (...) (2007:100)

Art and culture—as *Fairytale* has demonstrated—attract people, provoke public discourse and open up economic perspectives and prospects. There is nothing wrong with that. The deeper issue, however, is the present relationship between cultural production and an omnipresent global market system as well as economic thought, which appears to have become hegemonic and undermines what is actually valuable. The ramifying problems this brings are linked to an all-encompassing economic machinery as the driver of everything, and through which—according to Hardt and Negri—"capital not only brings together all the earth under its command but also creates, invests, and exploits social life in its entirety, ordering life according to the hierarchies of economic value" (2009:ix). A cartoon about the *Fairytale* project entitled "Art Market *extra*" illustrates these issues in subtle and illuminating ways by staging a world of pretence in which art is depicted as a morally depraved action field contaminated by unapologetic complicities and avaricious intents.[102]

From the ANT perspective, the good and the bad can be seen as embedded in the real, which means that the real is also embedded in the good and the bad (cf. Law 2009a:155). In this case study, ANT's

---

[102] Published in the *Neue Frankfurter Zeitung* the cartoon transforms public rumours about Documenta's *Fairytale* project into a sales talk about art. Wearing dark sunglasses and a red tie, Paul the cat—an art dealer and the owner of "Paul's Fine Quality Street Art"—talks about art with an unshaved bully and prospective buyer. The dealer offers what he calls a "new form of art", which is supposed to be available at Documenta—a sort of "briefcase sculpture" containing "1,002 Chinese brown rats", which, he claims, is more precisely a kind of "social, global big sculpture" of brown rats coming from all parts of the population that will take pictures and "scout your apartment". In duping the sceptical buyer, who can only see two rats, Paul asks: *Who are the true artists?* And gives the answer right away: *I have always said: the collectors!* The rats now decide to execute a previously hatched plan agreed with the "boss", namely to gnaw a hole through the briefcase and disappear.

Fig. 2.56

material semiotics has thus provided an understanding of both the ethical/political ambiguity of the real—and the ways in which basic ambiguous realities are manufactured and performed. The theory's assumption that these humanly constructed realities are not destiny but may be "remade" (Ibid.) brings us again to the pertinent question whether there might exist a *different* mode of "storytelling" for *Fairytale* to enact different realities as well as different relationalities. The question is what the project could have done to follow a different knowledge-building and socio-political strategy for potentially enlightening inquiries into its own framework of assumptions and how

Fig. 2.57

Fig. 2.59

Fig. 2.58

such commitments could foster truly social, ethical and "reflexive" cultural work practices.

As outlined earlier, *Fairytale* was based in the first place on the self-serving strategies of cultural and economic actors. Among them Ai Weiwei is the key stakeholder of everything and, according to Köppel-Yang, *Fairytale* advocates a "cliché" about Chinese culture (2007:95) despite being promoted in Kassel as an innovative artistic approach in its own right. To create social conditions in the name of art in order to learn in civilised and peaceful ways about other human beings or about human differences across ethnic/cultural lines is perhaps what this group of actors has achieved best. However, the lack of reflection on their own roles and public responsibilities as well as other forms of engagement that were missing prevented them from developing ideas that would have helped them not to be solely used to design a large art event, but instead to address other political/global challenges or urgent social issues.

To conclude, the *Fairytale* project seems to have perfectly matched what Dyer called the "ten-tonne truck that was Documenta 12" (2007:779). Its sheer material dimensions and ecological footprint must be considered as having negative societal value. Furthermore, the amount of more than 3 million euros needed to realise the project strengthens the impression that the *Fairytale* project might have been driven by economic incentives in the first place, and not by socially and culturally oriented epistemic deliberations as was claimed. This is clearly supported by the ANT analysis, which has opened up a different, much wider "networked" perspective on the agency of the project's key stakeholders. Thus, as *Fairytale*'s funding was generated almost immediately, the key interests and imaginations of its stakeholders may be seen as being associated with expectations concerning material gain as well as intentions to build up Ai Weiwei's position in the global art market in which China had started to play a significant role. From such a perspective, Documenta's decision to engage with potent stakeholders from outside its institutional structures and defined public responsibilities—among them a commercial art gallery representing Ai Weiwei's works globally—must be seen again as highly problematic.

As mentioned earlier, "Aid as Art" might have given *Fairytale* a more sustainable perspective, which is both social and political. However, as Dyer (2007:779) notes, the project was "loosely based around the fact that the Brothers Grimm compiled their fairytales in Kassel". Hence, it seems instead to have been designed to draw a tremendous amount of media interest because of the "tenuous conflation of this fact with the tradition of storytelling in China" (Ibid.). Yet, although criticised for being "eviscerated" by the media and over-instrumentalised as well as leaving an "odd taste" (Schmid 2007, no pagination),

Ai Weiwei's project stole the show in Kassel and made him one of the "most sought-after artists" for the curators of international exhibitions after Documenta (McDonald 2008:17). As Philip Tinari (2007:453) observes, his idea to "round up 1,001 Chinese people from the artist's sprawling, blog-mediated social network, give them matching clothes and luggage, fly them en bloc to Kassel, billet them on bamboo bunks in Ai-designed temporary quarters inside an old textile factory (...)" best describes how the *Fairytale* action plan contributed to the concept-driven and idiosyncratic apparatus of modernity.

Finally, the criticism of Actor-Network Theory that I want to discuss in relation to the *Fairytale*-project is that the theory addresses the local and contingent but pays little attention to the wider social (global) structures. One of the main criticisms of ANT, according to Walsham (1997:472-73), is the theory's focus on how things "get done" while the wider social and institutional macro-structures of society, which essentially shape socio-material practices, are excluded from its scope. My analysis of *Fairytale* as the outcome of cultural work by institutional and economic actors has, on the one hand, offered a way of looking at the social, political, epistemological and local aspects of cultural work. On the other hand, Actor-Network Theory has helped to include an appraisal of and reflection on the wider context of contingencies and conditions such as the global market system and its power structures, which make (local) cultural production and practices of various kinds vulnerable to disruption. Thus, the theory has opened up an awareness of interdependent and more complex (economic) macro-conditions in which cultural work is embedded. Moreover, if we see *Fairytale* again as an assemblage of socio-material-semiotic engagements and the "reality-work" (Law 2009b) of diverse actors—human and non-human—the macro-conditions in particular seem to have a negative impact on both knowledge-building endeavours on the local/social micro-level as well as practices of all sorts that today "make a more or less precarious reality" (Law 2009a:151).

In the three case studies, I have analysed key issues and presented insights concerning cultural work and the production and dissemination of knowledge. In chapters 3 and 4, I will draw on the conclusions of the case studies in order to rethink cultural work and sketch out possibilities of a practice-based "epistemology" for the post-industrial cultural workplace. It is clear that there are many challenges that must be discussed concerning the creation of "third" spaces of knowledge (Turnbull 2000) in which different uncertainties, unrecognised contingencies, the limitations of our knowledge as well as new learning dimensions of a mature knowledge society (cf. Felt, Wynne et al. 2007:63) could play a significant role. In order to embed these issues in what Hardt and Negri have conceptualised as the social and economic order of the common (2009:x), we must address the key challenges taking into account the relationalities in socio-epistemic and moral-ecological-political terms as well as governance issues for cultural work. Moreover, these challenges resonate with Beck's (1986) risk society narrative, which draws together the global situation, financial turmoil and climate change that confront our lives with new dilemmas and engender new risks. I will include these issues in my subsequent reflections on knowledge work in the post-industrial culture.

# III – Rᴇᴛʜɪɴᴋɪɴɢ Pᴏsᴛ-Iɴᴅᴜsᴛʀɪᴀʟ Cᴜʟᴛᴜʀᴀʟ Wᴏʀᴋ ᴀɴᴅ Cᴜʟᴛᴜʀᴀʟ Sᴘᴀᴄᴇs ᴏғ Kɴᴏᴡʟᴇᴅɢᴇ

> At the beginning of the twenty-first century, we see modern society with new eyes, and this birth of a "cosmopolitan vision" (Beck 2006) is among the unexpected phenomena out of which a still indeterminate world risk society is emerging. Henceforth, there are no merely local occurrences. All genuine threats have become global threats. The situation of every nation, every people, every religion, every class and every individual is also the result and cause of the human situation. The key point is that henceforth concern about the whole has become a task. (Beck 2009:19)

The aim of this chapter is to address issues that have thus far been identified as key challenges for the future of cultural work. As we have seen, the recognition of contingencies, our non-knowledge, (self-)reflexivity, route maps out of the demoralised terrain of neoliberalism as well as the (social) accumulation of the common vs. economic accumulation strike at the very heart of aspirations for development. Can we achieve such development in ways that support the creation of constructive linkages between new socio-epistemic practices, salient environmental issues and the knowledge capacities of diverse knowledge actors whose significance Felt, Wynne et al. outline (2007:15)?

As Felt, Wynne et al. point out, many of our contemporary doubts and uncertainties resonate with Beck's (1986) risk society narrative and cannot be separated from the now visible harmful consequences of technology—after decades of "profound collective emotional investment in science, as expected infallible producer of endless technical fixes to societal problems produced by human fallibility" (2007:13). Beck's term of the *risk society* epitomises an era of modern society that "no longer merely casts off traditional ways of life but rather wrestles with the side effects of successful modernization" (2009:8). In *World at Risk* (2009), Beck re-emphasises the "precarious biographies and inscrutable threats that affect everybody and against which nobody can adequately insure" (Ibid.). According to Beck, the world risk society must be seen as a "*non-knowledge society*" (emphasis in original) characterised by the global existence of incalculable global threats, numerous dilemmas and by our non-knowing in particular:

> Living in world risk society means living with ineradicable non-knowing (*Nichtwissen*), or, to be more precise, with the simultaneity of threats and non-knowing and the resulting political, social and moral paradoxes and dilemmas. Because of the global character of the threat, the need and burden of having to make life-and-death decisions increase with non-knowing. Talk of the "knowledge society" is a euphemism of the first modernity. World risk society is a *non*-knowledge society in a very precise sense. In contrast to the premodern era, it cannot be overcome by more and better knowledge, more and better science; rather precisely the opposite holds: it is the *product* of more and better science. Non-knowledge rules in the world risk society. Hence, living in the milieu of manufactured non-knowing means seeking unknown answers to questions that nobody can clearly formulate. (115)

While the role of science can be perceived as the motor of economic welfare, Beck's analysis outlining the key challenges of the risk society in terms of reshaping and recreating our present political and epistemic identities raises questions concerning alternatives to scientific knowledge. To reflect on these alternatives is, however, not the key concern of this book. Instead, I will address the issue of how to develop democratic and collective capacities as well as to incorporate uncertainties, reflexivity and knowledge-based doubt and reluctance (cf. Felt, Wynne et al. 2007:60) into cultural work practice. This seems reasonable to me, although not entirely beyond contestation from within the different ontological frameworks in which, as we have seen, cultural work, democratic values and interests are embedded

The three case studies have served to investigate cultural work from an empirical, ontological and socio-epistemic perspective. In an attempt to understand social, cultural, economic, educational, reflexive and ethical dimensions, the case studies have provoked thought in different areas of civic engagement in the humanities and the arts. The Gotthard Base Tunnel exhibition has illuminated the role and technocratic/reductionist functioning of public "dialogue" based on instrumental values and institutionalised commitments in politics and the economy. The moral-ecological engagement of *Manufactured Landscapes* appears to be at odds with the representation and late realisation of the compelling economic and ecological evidence that the present crisis invokes. In doing so, it raises questions of how to sensitise public awareness in offering knowledge of the multiple (global) aspects of environmental destruction. Finally, the case study of the *Fairytale* project at Documenta 12, which has uncovered a rather mixed record of achievements, has revealed how the project was linked to capitalist/economic frameworks and thus might have been based on key incentives to repatriate profit and strive for economic accumulation.

A central issue that runs throughout this book is the problem of (self-) reflexive thinking as well as the creation of spaces of knowledge for public reflection. The key questions are therefore what prospects exist for such spaces to engage citizens in processes of more basic, reflexive and rigorous thinking for acknowledging uncertainties, knowledge-based doubt and our non-knowledge of the world in particular. What sort of *reflexivity* is needed to sustain genuine contingency? These questions are important and need to be answered in relation to the larger institutional and perhaps cultural changes required to open up a much wider panorama of possibilities for the spatialisation and dissemination of knowledge as well as reflection on the principal role of public spaces of knowledge.

## (Self-)Reflexive Thinking in the Global Risk Society

In *World at Risk,* Beck provides a more subtle understanding of "reflexivity" and by defining it in relation to political and institutional commitments, Beck's analysis also provides the means and frameworks for the tasks of what we may consider to be "reflexive" cultural work. Referring to Anthony Giddens' and Scott Lash's (1994) view, Beck outlines (as mentioned above) that "reflexive" modernisation (in keeping with its literal meaning) is primarily associated with "knowledge (reflection) concerning the foundations, consequences and problems of modernization" (2009:119). In his view, reflexive modernisation *is* the result of "side effects of modernization" (Ibid.) in the first place. Furthermore, in acknowledging the difficulty of distinguishing between what we *know* and what we *do not know*—as knowing and non-knowing are always relative affairs—Beck notes:

> In the first case [Giddens' and Lash's view], one could speak of reflection (narrowly construed), in the second, of the *reflexivity* (in the wider sense) of modernization—in the wider sense because reflexivity, in addition to reflection (knowledge), also involves the idea of a "reflex" in the sense of the (preventive) effect of *not* knowing. (Ibid.)

Beck's distinction between reflection and "reflexivity" is a very productive one. Beck's view, which emphasises the salient dimensions of our non-knowledge, contains the notion of how the world and things must be treated from the perspective of our non-knowing and must be discussed in terms of what we *can* know about the world and what we should convey as "reflexive (non-)knowledge" to others. Beck's notion that what we normally claim to be our knowledge of the world is not the "medium" but instead "more or less reflexive" *non-knowledge* (122) thus becomes a central issue in the post-industrial culture with consequences for cultural work.

## A Second Modernity of (Self-)Uncertainty

Today we are facing an increasing number of incalculable threats, global uncertainties and risks with unknown consequences. Pertinent examples of unrecognised contingencies are the Fukushima Daiichi nuclear disaster in Japan in 2011 and the 2010 oil spill in the Gulf of Mexico.[103] These environmental catastrophes must be seen as examples of disturbing side effects of what Beck refers to as the *linear* modernisation process (125-26). Beck's view of a second *non-linear* modernity with more "consensual versus dissenting agent networks, queries, methods, guiding hypotheses, scenarios, assessments and evaluation of risks and threats" (126) is one we should welcome. There are, however, numerous theoretical as well as practical epistemic challenges lying ahead concerning the role of both epistemic actors themselves and rational/imaginative forms of knowledge and debate needed to sustain these challenges in the broad field of cultural production. As Felt, Wynne et al. point out, there is a broader social and intellectual involvement in how knowledge is manufactured and validated and the "co-production of science and society changes the very meaning of notions like objectivity and rationality" (2007:77). Yet, from the perspective of cultural work and the need for new impulses

---

[103] The worst oil spill in decades, which occured in April 2010, extended into precious shoreline habitat along the Gulf Coast. Emerging documents show that British Petroleum downplayed the possibility of a catastrophic accident at the offshore rig that exploded. British Petroleum suggested in a previously published environmental impact analysis that an accident leading to a giant crude oil spill as well as serious damage to beaches, fish and mammals was "unlikely, or virtually impossible". <http://www.npr.org/blogs/thetwo-way/2010/05/gulf_oil_spill_latest.html> downloaded 4 May 2010.

to anticipate the consequences of (rational) knowledge practices of different kinds, the core problem with rationality (and objectivity) is to find new ways of adopting *different* unfamiliar and educational rationalities. This evidently includes questions of non-knowing and new forms of greater *reflexivity* on the part of diverse actors and publics that should be able to respond to more rapid developments of different kinds, technoscientific progress in particular. Beck describes these issues as part of a greater range of challenges with regard to our monorational knowledge-building practices and epistemic sensitivities including the challenge to institutionally address the limitations of our knowledge that mark the change to the second modernity of (self-)uncertainty:

> [T]he foundations of the (economic, technical, political, scientific, etc.) monorationality characteristic of linear modernization, which is oblivious to consequences, begin to crumble. This very monorationality is still advocated today in the guise of systems theory (with the suggestion that functionality and autonomy depend on screening out the external perspective). Both factors—the question of *our own* inability-to-know and the ability to adopt the perspective of alien rationalities—mark the transition to the second modernity of (self-)uncertainty that is simultaneously manufactured by civilization and known. Only then can we pose the general question of how these antagonisms and differences of *known non-knowledge* can be inter-connected, played out and combined into decision-making procedures in new forms and forums. (2009:126)

These wider issues have yet to be clearly recognised. But the major problem from the perspective of cultural work is the development of social/intellectual capacities in order to agree on collective uncertainties as well as unrecognised contingencies that challenge human (self-) reflexivity and (un)learning. The task for cultural work is thus to open a more intense dialogue about these issues in order to create significant civic engagements. As Felt, Wynne et al. outline, the way to achieve this is through "new institutions and procedures for more inclusive pluralistic discussions, learning, and challenge" (2007:82). For his part, Banks points to the virtues of alternative cultural production to enhance possibilities for "reflective self-comprehension and social action" (2007:164; cf. Willis 2005:76). He sees certain yet to be elaborated potentials for cultural work, noting that

> [a]lternative cultural production (...) must in some way be credited for opening up possibilities for self-reflexivity, for engendering more creative and autonomous attitudes and for (at least potentially) furnishing possibilities for progressive social practices—even as these possibilities remain vulnerable to the incursions of commodification and marketization. (Ibid.)

To conclude, cultural work should not be judged for lacking *a priori* progressive or transformative potential. Cultural work should be seen as being linked, at least potentially, to positive and constructive socio-epistemic, socio-political and ethical dimensions that typically remain close beneath the surface of what we normally consider to be "cultural work" and "cultural workers" with important knowledge capacities and concerns that often remain unarticulated. This brings me again to the conclusion that to acknowledge the foundational contingencies and non-knowledge as well as to expose various kinds of uncertainty—in scientific knowledge and the self, for example—constitutes a crucial element of what I consider to be *knowledge work* in the post-industrial culture. Moreover, open-ended dimensions of collective learning and unlearning should become particularly important in the framework of practical commitments to democratic values that are negotiated on a broader basis and to interests that are shared by many. These dimensions through which cultural work (cultural workers) may contribute to a more reflexive and collective awareness seem to have been reduced to ideas of instrumental learning only, as documented in this book. In addition, such new learning dimensions are particularly important in a world in which threats and risks of all kinds have become world-sized problems "beyond status and class" that ultimately affect everybody (Beck 2009:22).

So far, I have been discussing different theoretical approaches in order to frame cultural work and the dissemination of knowledge in the post-industrial culture. These approaches have helped to open a much wider panorama of possibilities for cultural work to shape society from within the humanities and the arts. In the next section, I shall examine closely some of the key points in Banks' (2007) discussion of the prospects of post-capitalist cultural work.

# Post-Capitalist Cultural Work

A central insight emerging from Banks' summary of the route maps from the demoralised terrain of neoliberalism criticised by Beck, Gorz and Harvey is that striving towards instrumental ends has "most effectively diminished moral guarantors of selfhood, work and sociability historically associated with traditional life" (2007:168). By taking into consideration the deep antagonism towards capitalist notions of progress that these critics share, for Banks the "seismic shifts that would require a wholesale ethical retooling of societies" advocated by Beck and Gorz are, however, a "big ask" (Ibid.:169). It is indeed hard to say to what extent relevant actors may sustainably combine the worlds of economic and non-economic moral values, which may already interrelate in contexts where cultural work has a positive impact with far-reaching implications for society. By validating analyses such as those of Beck, Gorz and Harvey, Banks acknowledges that these issues are at stake when he concludes that

> [b]uoyed by their creative spirit [Beck, Gorz and Harvey's], but aware of my own imaginative shortcomings, my remaining intention is simply to address how far existing economies may *already contain* diverse examples of the progressive (utopian) interplay of economic and non-economic moral values in the contexts of cultural work. (170)

For Banks, anti-capitalist virtues such as bartering, the invention of art currencies, digital gift economies (of which file-sharing is an example), or the "bedroom culture" of music producers, digital artists as well as film-makers etc. (171-78) are examples of new convergences that could support—at least potentially—the rise of new models of social exchange and alternative non-capitalist economies on a broader scale. However, these alternative economies do not show the emergence of new regimes of what we may consider to be genuine post-capitalist cultural work. Instead, they seem to continue to rely on conventional capitalist frameworks and systems of exchange. According to Banks,

> alternative currencies and the bartering or gifting of art and cultural goods remain marginal forms of economising, and ones that (as yet) appear ill-equipped to serve the more substantial and complex social needs of community members. Indeed, it is notable that despite radical intent, proponents of alternative economies often remain dependent upon the main-stream capitalistic economy for a whole variety of resources—ones as yet unobtainable within a non-capitalist economic framework. For example (...) artists themselves often rely upon conventional "second" jobs in order to survive. Additionally, the production of alternative currencies may well rely upon resources obtained through fiat money purchases or conventional systems of commodity exchange. (179)

To conclude, Banks sees only minor possibilities for cultural work to build "miniature democracies" in order to generate "some *collective* vision that can progressively transform the character of economic life" (186). What is needed is, to my mind, a major emphasis on both the political rethinking and democratic reshaping of cultural work. It should appear as a central preoccupation instead of reconfiguring the art-commerce relation (185) under imagined harmonising market conditions under which the present hegemonic capitalist frameworks of production might begin to crumble. The task is to rethink cultural work from the perspective of its democratic capacities as well as the human capacities for collective (self-)reflexivity, perception and (un)learning as outlined earlier. These issues could be promoted together with the diverse civic "knowledge-abilities" (Felt, Wynne et al. 2007:10) that socially dispersed knowledge actors (Ibid.:15) and cultural workers can bring into play. Moreover, these issues resonate with Hardt and Negri's concept of biopolitical production, which they have explored in both *Commonwealth* (2009) and *Empire* (2000). Based on the idea of the biopolitical turn of the economy I will argue in the next section that Foucault's concept of biopower together with Hardt and Negri's notion of the (social) accumulation of the common may put cultural work into perspective with regard to the many urgent issues that need to be addressed in the post-industrial culture.

## Cultural Work and Biopolitical Production

According to Hardt and Negri, there are three major current trends in how labour is transformed, namely through the hegemony/prevalence of "immaterial production" in the processes of capitalist valorisation (2009:132), the technical composition of labour in the feminisation of work (Ibid.:133) and the technical composition of labour resulting from new patterns of migration and processes of social and racial mixture (134). *Immaterial* production is seen as a key characteristic of new production and, according to Hardt and Negri, the biopolitical nature of labour is constituted by immaterial factors and goods or in other words by what these authors call the "labor of the head and heart":

> Images, information, knowledge affects, codes, and social relationships, for example, are coming to outweigh material commodities or the material aspects of commodities in the capitalist valorization process. This means, of course, not that the production of material goods, such as automobiles and steel, is disappearing or even declining in quantity but rather that their value is increasingly dependent on and subordinated to immaterial factors and goods. The forms of labor that produce these immaterial factors and goods (or the immaterial aspects of material goods) can be called colloquially the labor of the head and heart, including forms of service work, affective labor, and cognitive labor, although we should not be misled by these conventional synecdoches: cognitive and affective labor is not isolated to specific organs but engages the entire body and mind together. (...) What is common to these different forms of labor, once we abstract from their

> concrete differences, is best expressed by their biopolitical
> character. (132)

The idea of cultural work as a union of affective and cognitive capacities as well as moral and intellectual imaginations that develop from the labour of the head and heart is seen as central to the ability of *knowledge work* to convincingly address and promote salient issues at stake in a democratically committed post-industrial knowledge society. New ways should be found to promote the much wider range of cognitive virtues and deeper social values, as claimed by Armstrong (2009, no pagination). Yet, what I see as an as yet unarticulated key challenge to cultural work is to come to terms with both the promotion of agency and collective awareness as well as a deep positive contemplation of life. This brings us to Foucault's concept of biopower.

According to Hardt and Negri, Foucault's concept is primarily focused on the "power over life" and the power to "administer and produce life—that functions through the government of populations, managing their health, reproductive capacities, and so forth" (2009:57). Hardt and Negri further argue that there is a "minor current" in Foucault's analysis of biopower which sees life expressed in forms of resistance and freedom. In other words, this means that there exists another dimension of the power of life that "strives toward an alternative existence" (Ibid.). According to Hardt and Negri:

> [Foucault's] analyses of biopower are aimed not merely
> at an empirical description of how power works for and
> through subjects but also at the potential for the production
> of alternative subjectivities, thus designating a distinction
> between qualitatively different forms of power. (...) When

one defines the exercise of power as a mode of action upon
the actions of others, when one characterises these actions
by the government of men by other men—in the broadest
sense of the term—one includes an important element: freedom. (...) Biopolitics appears in this light as an event, or
really, as a tightly woven fabric of events of freedom. (59)

*Biopolitical events* can thus be seen as individual acts of freedom and
resistance. These are crucially important for Hardt and Negri for the "production of life" (61) while, at the same time, they constitute an important
element of the biopolitical turn of the economy. Their claim that "biopolitical production puts bios to work without consuming it" (283) may further
be seen as a relevant stance from which the truly endangered ecological
complexity of the world and the common wealth that we all share (viii)
may resonate with (self-)reflexive, ethical and life-sustaining collective
aspirations. In promoting an open and democratic dialogue about new
ways of collective engagements to accumulate the common, the focus on
our common wealth, I suggest, should constitute an essential element of
cultural work. The idea of producing wealth without consuming beyond
both our planetary limits and diminishing the resources on a finite
planet is, according to Hardt and Negri, a crucial element of biopolitical
production. They argue that

> [a]ccumulation of the common means not so much that we
> have more ideas, more images, more affects, and so forth
> but, more important, that our powers and senses increase:
> our powers to think, to feel, to see, to relate to one another,
> to love. In terms closer to those of economics, then, this
> growth involves both an increasing stock of the common
> accessible in society and also an increased productive

> capacity based on the common. One of the facts that make
> us rethink such concepts of political economy in social
> terms is that biopolitical production is not constrained by
> the logic of scarcity. It has the unique characteristic that it
> does not destroy or diminish the raw materials from which
> it produces wealth. (283-84)

This section has focused on the biopolitical dimensions of cultural work. As we have seen, there is a link between cultural work and the (social) accumulation of the common as well as strategies that could seek more resilient long-term and anticipating social impact. While diverse knowledge actors and knowledge capacities can play a significant roles in the broad field of immaterial production, the significance of cognitive and emotional capacities as well as the social/collective dimensions of our existence need to be acknowledged more thoroughly. According to Hardt and Negri, these issues outweigh material commodities or the material aspects of commodities in the capitalist valorisation process (132) and are thus of *greater* relevance.

## Conclusions

In this chapter, the key challenges to *knowledge work* in the post-industrial culture have been addressed. The chapter has attempted to address these challenges from a much wider horizon of possibilities for cultural work. This includes tensions and contradictions which have been identified. The chapter has provided ideas as to how cultural work could become concerned about the whole, which for Beck has become a task (2009:19) necessary for dealing with collective threats and which addresses our non-knowledge, ontological uncertainty, foundational contingencies and dilemmas which are linked to deeply human, social, political and ethical concerns. As we are now facing the harmful impacts and accumulated effects of our encroaching civilisation, altered lands-capes, burgeoning industry and global warming etc., rethinking cultural work means becoming actively and democratically involved in reinventing processes of knowledge production and dissemination. Furthermore, in order to experiment with present and new forms of civic engagement, (self-)reflexivity and new ways of collective (un)learning, *knowledge work* in the post-industrial and globalised culture should be seen as a challenge for cultivating new dimensions of learning and imagination. These are collective issues with implications for governance and the general public as outlined in the conclusions to the case study of the Gotthard Base Tunnel exhibition (Figure 2.11). If the social accumulation of the common and cultural work may be expected to consolidate in the context of new conditions, purposes and questions that sustain projected collective futures, we need achievable and realistic change now. "Business as usual" should no longer be considered the only contending option for cultural work. This will mean changes in how cultural work is done, what cultural work is engaged in and how cultural work and knowledge are synthesised to what we need to understand or change by contributing to the public and global good.

In chapter 4, I will sketch a rationale for an "epistemology" of cultural work which I see linked to practices of engagement by which both knowledge and space are constituted (cf. Massey 2005). The aim of chapter 4 is to provide practical support for practices of mediation, inter-pretation, representation and education prompted by civic intention. In addition, these practices should encourage self-inquiry at the most perso-nal level as well as create new forms of dialogue between the humanities, the arts and the public. These are not simply different "forms" of humanities- and arts-based civic dialogue, but practices associated with the knowledge required to implement political, epistemological, aesthetic, ethical and democratic concerns which need to focus on the neglected or unknown and thus unpredicted effects on society in risk modernity (cf. Felt, Wynne et al. 2007:11). Chapter 4 will also offer insights into how language, reflexivity and temporality may play a role as practical tools to construct and perform knowledge and realities, which also involves the importance of recognising the contingencies which underlie knowledge. According to Felt, Wynne et al. "[a]wareness (...) of the limitations of our knowledge, thus modesty in the claims we make with it, is another crucial learning dimension of a mature knowledge society" (Ibid.:63). These are the real "public" issues of *greater* significance that must be addressed and thus given a political space. In doing so, the rejection of cultural workers of market-led capitalism might simultaneously pave the way for the emergence of a more anticipatory dynamism as well as coming to terms with legitimate civic expectations of democratic deliberation and accountability.

# IV – Epistemology of Post-Industrial Cultural Work

> If space is genuinely the sphere of multiplicity, if it is a realm of multiple trajectories, then there will be multiplicities too of imaginations, theorisations, understandings, meanings. (Massey 2005:89)[104]

Massey's notion of space as a "realm of multiple trajectories", which suggests that cultural work is a constructively engaged spatial form of work, converges nicely with Henri Lefebvre's concept of "spatial practice", that is, the (mental) projection onto a "(spatial) field of all aspects, elements and moments of social practice" (1974; 2009:8). According to Lefebvre, we are "confronted today by an indefinite multitude of spaces, each one piled upon, or perhaps contained within, the next: geographical, economic, demographic, sociological, ecological, political, commercial, national, continental, global" (Ibid.). The use of these spaces for democratic deliberation must be seen as a challenge to design distinct forms of social/epistemic interaction as well as to create new forms of public participation. Yet, how should this be done in both theoretical and practical terms? The starting point for new practices for cultural work and social action is a socio-epistemological approach which will be highlighted in this chapter. This approach offers, on the one hand, a fundamentally different understanding of collective practice; on the other hand, it opens a new horizon from which alternative potential knowledge as well as ambiguities, contingencies, uncertainties and risks can be addressed (cf. Felt, Wynne et al. 2007:35).

---

[104] Massey's notion of the "multiplicity of trajectories" (2005:119), which conceives space as an imagined open sphere, implies the "social in the widest sense: the challenge of our constitutive interrelatedness; the radical contemporaneity of an ongoing multiplicity of others, human and non-human" (Ibid.:195).

## A Socio-Epistemological-Political Model of Learning

How can we develop a different culture of governance capable of bringing these issues into play and, at the same time, come to terms with the many real and serious problems that we are facing? Heinz von Foerster's socio-epistemological model of learning opens up a thought-provoking and practical perspective from which solutions to these problems can be designed. Marcelo Pakman's[105] "organizing elements" (Pakman 2003:114) provide additional support for practices and potential solutions from within alternative second-order frames in ethics, aesthetics, pragmatics and politics.[106] Finally, this chapter aims to provide a methodological and political anchor in order to address key challenges and what I consider to be key qualities of post-industrial cultural work (Figure 4.1). These issues strongly

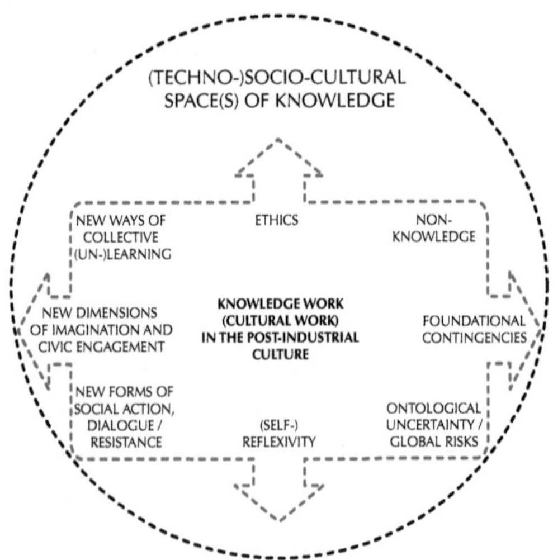


Fig. 4.1



---

[105] Marcelo Pakman is a community psychiatrist and systemic family therapist.
[106] Pakman's "organizing elements" build on von Foerster's Second-Order Cybernetics—the "Cybernetics of Cybernetics"—, which emphasises the central capacity of human beings to observe, thus theorising them as observing systems (2003:107).

resonate with the idea of the spatial characteristics of all earthly/worldly phenomena. Combined, Pakman's "poetics" in therapeutic conversations, von Foerster's (1973; 1998) trope of "reality equals community", and Donald Schön's (1983) concept of "knowledge-in-action" provide the skeleton for a structured socio-epistemological-political approach to the production and dissemination of knowledge in the post-industrial culture. In this triad, Pakman's "poetics", which serves to channel participation, works as an "implicit articulated series of organizing elements of therapeutic conversations" (Pakman 2003:113). More importantly, in his model the psychotherapist does not play the role of a "privileged interpreter", but, instead, introduces reflexive strategies in order to "navigate multiple (...) interpretations, embedded in the actions, perceptions (...) that are the matter of social, psychological (...) human life" (Ibid.:111). It is this therapeutic

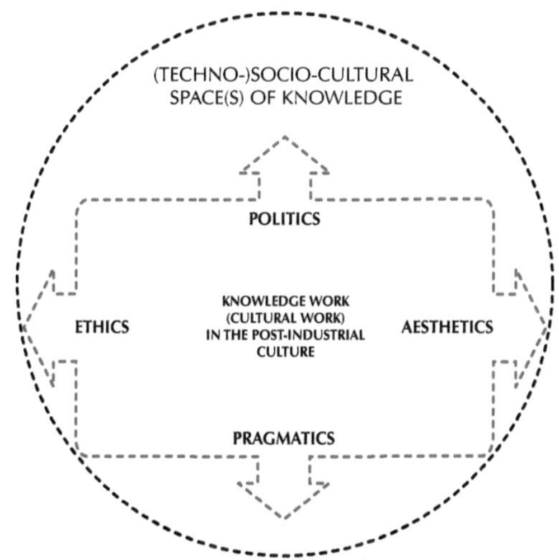


Fig. 4.2

approach which is oriented towards creating ethical, aesthetic, pragmatic and political convergences that could play a principal as well as constitutive role in cultural work (Figure 4.2). In addition, the "poetics" allows us to manage diverse challenges (as I will show) and to reflect on perspectives and realities from a second-order point of view.[107] In the following sections, I will explore how ethics, aesthetics, pragmatics and politics as well as language, reflexivity and temporality fit into cultural work as methodological second-order positions to accommodate the many ambivalent and often questionable voices of the humanities and the arts.

[107] According to Pakman, the "poetics" is based on "*theories-in-use*, which are *theories of action*, structured, patterned ways of moving about in certain types of situations in order to achieve certain specific goals [Schön 1983]". These poetic elements, he further notes, can be described as an "interconnected network of concepts of a level of abstraction close to practice, with more or less identifiable roots in many different sources of learning" which serve in general to "capture a possible understanding of the understandings that mediate professional actions, of our 'knowledge-in-action'" (2003:113).

# Ethics

Ethics must be regarded as an essential and necessary requirement for human life. It is deeply connected to questions such as "What do we do?", "How should we decide?" and "How should we act?". Ethics is also a method for defining and practising values. Today, ethics is, however, often seen as a matter of "expert analytical derivation and discovery rather than of collective, principled, reasoned reflection" (cf. Felt, Wynne et al. 2007:17). As a consequence, ethics is frequently treated as a mode of expert elucidation and knowledge (Ibid.:46). The proposed key to linking cultural work to such underlying ethical dimensions is von Foerster's *Ethical Imperative,* "Act always so as to increase the number of choices". This imperative supports us in creating a second-order perspective from which choice and human freedom appear as requirements that open up the possibility of "describing different ways in which things (...) could come to actually happen", including possibilities for new "variations and choices within restrictions" (cf. Pakman 2003:116). The following questions are posed in order to multiply alternatives and to address needs and priorities as well as to question established imaginations:

- What alternatives exist to create public (techno-)socio-cultural spaces of knowledge for cultural projects such as an exhibition, a conference, a film, a performance etc. on a given civic issue?
- What other alternatives and potential ideas were at some point available, but for whatever reason were not followed?
- Is there a way to reinstate any of those alternatives to replace current ones?
- How can these alternatives be conceived of as desirable possibilities to highlight potential commitments that could help to address societal problems or dilemmas?

- How can they be described as explicitly as possible and as potentially positive knowledge trajectories?
- How can they be transformed into new choices and commitments?
- How can they be used to reshape collective needs and "real" issues at stake?
- How can they be linked to ecological practices?
- How should these alternatives be conceptualised with regard to improving language and narrative?
- How could other educators, curators, change agents or cultural actors as well as audiences be involved so that they can provide expertise and alternative views that would help to increase collective knowing?

## Aesthetics

Ethics and aesthetics form a unity directed towards orientation, learning and action. This entanglement of both ethics and aesthetics is emphasised by Pakman, who suggests that

> [a]esthetics is here ethics as well. It is a choice among choices: the choice of those options we would prefer to see happening. No choice is complete until we aesthetically choose which ones we want to see happening. (...) [E]thics is always pre-announcing aesthetics, because there is a chain of choices leading to what we prefer to see happening, and orienting our actions and the learning we need to go through in order to act efficiently. (2003:117)

Von Foerster's *Aesthetic Imperative,* "If you desire to see, learn how to act" is introduced here as a potentially constructive and strategic position. This imperative calls for greater reflexivity on what we may learn when we realise preferred choices or, more precisely, which of the "identified options (...) we choose to *see,* and what (...) we need to learn in order to see them happening" (2003:116). How can we come to terms with the criteria that turn those choices into socio-political knowledge trajectories/spaces and thus prepare us for the management of more complex situations, societal problems and risks? By asking ourselves the following questions, we might be able to frame the basic intentions and goals that cultural work is associated with:

- What motivates the different actors to construct spaces of knowledge for public reflection and what are the values that guide them?
- What are the principles or criteria that are applied to make (techno-)

socio-cultural spaces of knowledge the preferred ones?

- What makes some of those spaces more useful to address a given civic issue, realise a project, or articulate interests and imaginations?
- What conflicts based on particular choices or preferred solutions could arise in these spaces?

If cultural work is the means by which choices can be realised and made effective, then these questions may help us to come to terms with new potential learning dimensions. They are able to do so since they call our attention to the knowledge claims of different actors as well as to the political dimension of our knowledge and new forms of democratic deliberation. By reshaping our choices we will be reshaping cultural work. These issues are entangled in the pragmatic dimension of how to *engage* through concrete action and to develop an understanding of the problems at stake.

## Pragmatics

If we accept that ethics and aesthetics support us in becoming more sensitised to the neglected aspects of learning and acting, then cultural work should be seen as being entangled in the pragmatic challenge of creating new learning environments as well as envisaging horizons for possible action. The following questions help to focus on the pragmatic dimensions of learning and acting, underlying choices, technosocial commitments and the wider socio-economic-political realities embedding our lives (cf. McCarthy 1996):

- What do cultural workers need to undertake in order to realise the preferred choices and turn them into knowledge trajectories for civic reflection?
- What questions are missing with regard to constructing these knowledge trajectories based on those choices?
- How should cultural work and concrete action be linked in order to move in the desired direction?
- What is missing in order to move in that direction?
- Does starting to move in a chosen direction mean that aspects of learning are augmented and multiplied? Or does it mean that important dimensions of learning are obstructed (cf. Felt, Wynne et al. 2007:70)?

Any attempt to rearrange cultural work is first and foremost a political challenge, a challenge confronting normative and societal forces such as forms of technical and political control as well as the challenge of the power of the market (as we have seen). This brings us to the (micro-)political dimensions of engagement and acting.

# Politics

Today, trust and confidence in politicians, political parties and institutions have disintegrated. Space is opening up for populists and demagogues. We therefore not only need, I suggest, to prepare cultural work for a new form of more explicit "micro-politics" (Pakman 2003:119), but we also need a (micro-)political context in which "ethics, aesthetics, and pragmatics occurs" and becomes effective (Ibid.). This means that to deal effectively with the many problems that we are facing, cultural work should be linked to understandable political goals built around matters of public concern. Furthermore, Massey's claim regarding the multiplicity of human/non-human convergences and spaces needed for public deliberation and democratic practice could be seen as a guiding principle in the process of negotiating these spaces from within the frameworks of power and control. Are these not issues that are manifest in the deepest aspirations and concerns of cultural work? Questions such as the following might sharpen our awareness of larger societal and political concerns as well as of the prevailing conditions of power and control:

- What micro-political strategies should cultural workers pursue in order to resist power and control?
- What political vision of societal needs is required in order to move toward "desired options"?
- How can those options be realised and made (politically) effective?
- At what level and in what direction should cultural workers start to act in order to realise those options?
- How should cultural work be (politically) positioned in order to promote collective learning and unlearning in the best possible way?
- Who should be involved and what interaction among different (political) actors is intended?

In Pakman's "poetics" there is no predefined concept or chronological order that would determine how to engage with the ethical, aesthetic, pragmatic and political aspects of social experience. He embeds the four positions in a networked whole of questions which I have transposed into the realm of cultural work and elaborated in the previous sections. The central aim of both Pakman's "poetics" and cultural work is to identify new options and make the choices we see much more explicit so that they may lead to (micro-)political actions. As Pakman notes, to "operate" from within micro-political premises and make these actions finally happen means to bring our awareness to ethics, aesthetics and pragmatics by "reflectively review[ing] the context in which interpretive practices happen, are born, promoted or maintained" (2003:111; 119-20). Ethics, aesthetics, pragmatics and politics can thus be seen as potentially constructive elements to culturally produce and disseminate knowledge. Further resources for rethinking cultural work and the knowledge required for acting are provided by the use of tools such as language, reflexivity and temporality, which I will explore in the following sections (Figure 4.3).

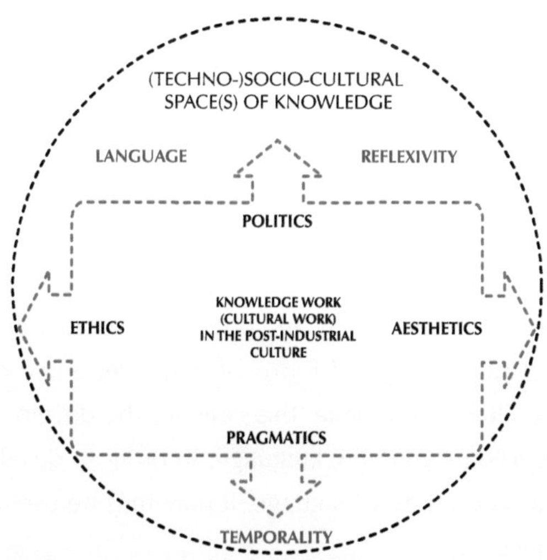


Fig. 4.3

# Language

Language embodies a wide range of practices such as translation and mediation, let alone the many mainstream forms of scientific representations in techno-hyped spaces which are communicated through diverse forms of language. Language, reflexivity and temporality may thus be seen as tools by which the "actual conversation" is steered (Pakman 2003:120). Language itself can further be seen as a participatory activity which plays a central role in (re)shaping both traditional forms of learning and more complex learning and communication processes (cf. Felt, Wynne et al. 2007:55). In Pakman's "poetics" language is seen as the key factor of human communication:

> We invent choices we are responsible for; we choose the options we find more desirable; we design actions to make our choices happen; and we try to operate at a micro-political level to make choices, and to initiate the actions to make them happen. (...) We do (...) things from within the systems in which we find ourselves operating or we design in order to operate. And we do it always using a language that captures this constant speaking as a part of the world we describe, design, suffer, enjoy, and act upon. (2003:120)

Given that language and practices of representation cannot be separated, using language to design new and open forms of learning could form the basis for new concepts and forms of dialogue. Since cultural work is strongly committed to dialogue, this calls for the development of new communicative idioms and new language. In order to develop (political) strategies based on such new language, it may thus be useful to consider the following questions:

- Which are the issues of greatest social need?
- How should these issues be addressed through public dialogue and deliberation?
- What areas of deliberation and debate should be fostered? What are those which are off limits?
- How engaged do cultural workers want to be with regard to those areas of debate and deliberation? What risks are they willing to take?

What we may call genuine participatory language is certainly language that is capable of keeping democratic engagement alive. Catalysing and provocative new language as well as new qualities of creative thinking of no *"either/or"*, but rather a definite *"both/and"*, as Gogan (2002:32) argues, may thus be seen as a constructive constituent to transform knowledge into new learning dimensions. This brings us to the elementary tool of *reflexivity*, through which deliberation as a fundamentally interactive activity constitutes another important second-order stance to sustain the more reflexive capacities of cultural work.

# Reflexivity

The development of a proper culture of deliberate reflexive cultural work practice may be seen as fundamental to both engaging in learning as well as reinforcing learning experiences. But also what Felt, Wynne et al. call "(self-)*reflective*" reasoning or what they call "indirect learning, as distinct from instrumental, direct learning alone" (2007:65), means that *(self-)reflexivity* holds its own distinctive place as a fundamentally human ability. Adair Linn Nagata defines self-reflexivity—for the purposes of intercultural communication—as "having an ongoing conversation with one's whole self about what one is experiencing as one is experiencing it" (2004:160). She argues that

> [t]o be self-reflexive is to engage in [a] meta-level of feeling and thought while being in the moment. The strength of being reflexive is that we can make the quality of our relationships better at that time in that encounter (...). (Ibid.:140-41)

Yet, (self-)reflexivity—understood as a circular relationship and capacity of an individual person to recognise or respond to forces of socialisation—reaches beyond the dimensions of an ongoing *reflexive* conversation with one's whole self about what one is experiencing.

## Reflexive Understandings and Interactions

It follows from Nagata's view and von Foerster's notion that human beings see themselves "through the eyes of the other" (von Foerster 1984),[108] that reflexivity is more than an "isolated introspective exercise, which would be necessarily blind to the interactive nature of our actions and the effects of our interventions (...) and languages", as Pakman notes (2003:121). Reflexivity, Pakman argues, is about "mutual observation, multilateral decision making negotiation". It opens up a new horizon from which we can design our actions and define the "quality of every representation" (Ibid.), and entails reflexive *understandings* and *interactions* within a framework of new options at hand. In Pakman's view:

> [I]ncreasing alternatives, choosing the more desirable ones, moving toward making them happen, and acting politically to operate more effectively at every step of the process, involves always using those same ethical, aesthetical, pragmatic, and political parameters to see ourselves and invite to see [sic] (...) all those choices and actions. In turn, it is this very reflective stance what [sic] makes all of those poetic elements to be [sic] mobilized into creating more options, choices, and actions. (Ibid.)

Such reflexivity may have an enlightening quality when it is directed towards unquestioned assumptions and practices in order to anticipate the consequences and effects of these practices and assumptions. Strate-

---

[108] Von Foerster's observation correlates with the Russian philosopher and theorist Mikhail Bakhtin's dialogic theory of communication that the self can only view itself from the perspective of the other. As Clark and Holquist point out, Bakhtin "conceives of otherness as the ground of all existence and of dialogue as the primal structure of any particular existence, representing a constant exchange between what is already and what is not yet" (1984:65).

gic thinking and greater reflexivity with regard to constructing (techno-) socio-cultural spaces of knowledge oriented towards disturbing realities, risks and contingencies could be cultivated by considering the following questions:

- What (techno-)socio-cultural spaces of knowledge are needed to create positive social and long-term ecological impact?
- What spatial structures are required to build such spaces in order to engage knowledge with immediate and longer-term impact?
- What (techno-)socio-cultural spaces are needed to foster (micro-)political strategic policy practices?
- What (techno-)socio-cultural spaces are required to mobilise other actors and short-term measures in the best way possible?

The answers to these questions, I would suggest, need to be linked to clear political ambitions and goals. These questions could be raised in order to find answers, yet this does not mean that cultural workers can relax and stand apart from micro-politics. Even that would be a micro-political act, with micro-political consequences. This is to say that these questions also feed into basic challenges such as those of the current dominant idioms of control that may obstruct important dimensions of (un)learning as well as challenges of how to publicly perform knowledge generally. This brings us to von Foerster's "logic of 'becoming'" (Pakman 2003:121), which is based on the concept of temporality as the final step to develop new knowledge trajectories in a socio-epistemological system of mutually reinforcing ethical, aesthetic, pragmatic and political linkages.

# Temporality

If we replace the verb "to be" with "to become" (cf. Pakman 2003:122), the perspective on our being-in-world (and the world itself) changes. Moving to an ontology which is based on a logic of "becoming" (von Foerster 1984) means acknowledging the changeable, open-ended and "fluid" dimensions of our existence and our experiences. Thus, temporality introduced here as a tool may not only bring cultural work to the level of a different awareness of the dimension of time, but equally emphasises the need of a durational process in order to bring both civic participation and constructive interaction into play. In Pakman's view:

> The events that make for the experience of the social actors in the situations we deal with and we find ourselves [sic], the attributions made during conversations, the qualities described and assumed to be constitutive of state [sic] of affairs, and the evaluations made of options and actions, their desirability, and the contexts in which they occur, all are seen from a different perspective (...). (2003:121-22)

To conclude, ethics, aesthetics, pragmatics and politics as well as language, reflexivity and temporality may be seen as potentially constructive elements to build resilient spaces of knowledge. In addition, the proposed second-order framework could provide the necessary basis to encourage cultural work to reflexively and more efficiently inquire into our knowledge, contingencies and certainties as well as new dimensions of (un)learning at stake—a requirement that is urgently needed today.

## Conclusions

In scientific methodology and social theory, cybernetics has been widely recognised as a conceptual tool. Many of its principles "have come to be taken for granted" and today, cybernetics thus contributes to "larger epistemological transformations" (Shanken 2003:19). Yet, is cybernetics and in particular second-order cybernetics a valuable methodological tool for the thorough appraisal and reconceptualisation of cultural work as discussed in this chapter? Is it possible to develop a resilient moral-political-ecological anchor for cultural work from within the framework of second-order cybernetics? According to Ascott, second-order cybernetics opens up a new understanding of both the concept of interaction and creativity as the discipline is characterised by the features of "feedback, self-determination, interaction, and collaborative creativity" (1990; 2003:242). He argues:

> [C]ontrary to the rather rigid determinism and positivism that have shaped society since the Enlightenment (...) these features (...) accommodate notions of uncertainty, chaos, autopoiesis, contingency, and (...) [a] view of a world in which the observer and observed, creator and viewer, are inextricably linked in the process of making reality—all our many separate realities interacting, colliding, reforming, and resonating within the telematic noo-sphere of the planet." (Ibid.:242-43)

One way of complying with von Foerster's and Pakman's second-order perspective on learning is to say that it provides us with a valuable, even quintessential bird's-eye-view—a true perspective "from above". This is surely an important requirement in order to reflect from a more "distanced" view on how problems can be solved, decisions can be taken or communication and collaboration can be defined. Furthermore, I suggest that the second-order approach is capable of promoting a more interactive and anticipatory dynamism based on reflexive inquiries into the construction of longer-term impact and the production of beneficial outcomes. To anticipate conditions in the real world and to teach the humanities and the arts to become better at engaging beyond their lateral territories should be seen as key challenges—and to design new learning dimensions from a socio-epistemological second-order point of view may become the starting point for new solutions and efficacy.

In my final conclusions in chapter 5, I will address some of the ongoing perspectives of this research and focus on those areas of knowledge work in the post-industrial culture in which learning may be most important.

# V– Conclusions and Outlook

The essence of knowledge is, having it, to apply it; not having it, to confess your ignorance. (Confucius)

In this book, I have tried to show how in-depth inquiries into the nature, practices and theory of post-industrial cultural work offer insights concerning cultural work and governance. I have argued that the emphasis put on uncertainty-oriented approaches and (self-)reflexive inquiries into our knowledge and, in particular, our non-knowledge pose a new challenge for cultural work. Yet, the practicability of these issues that have been problematised from different angles in various parts of my text remains vulnerable to disruption. Although people may share concerns over "risk" or "uncertainty", there exists, on the one hand, a tendency of humans to externalise contingencies and favourise certainties over uncertainties. On the other hand, obstructing important dimensions of learning as a result of efforts to sustain the predominant culture of instrumentalism poses, as we have seen, a core problem for both cultural work and the underlying technocratic and reductionist governance culture itself (cf. Felt, Wynne et al. 2007). According to Turnbull, it is only through critical reflection and the explicit articulation of our Western epistemological and political high stakes as to what we count as knowledge (cf. Turnbull 2000) as well as new forms of reasoning that we can learn to grapple with the rationale and authority of our knowledge and unrecognised dimensions of learning. But how should cultural work be effectively positioned with regard to these most salient issues? How could public trust and dialogue be created, and the social diffusion of diverse knowledge be managed? How can we address moral-ecological issues that are intrinsically linked to post-industrial socio-technical reality, and can this become a matter of routine in public deliberation? Is there a way out of our ecological dilemma, which—figuratively—can best be described as that of a mouse staring into the throat of a snake?

The three case studies presented in this book suggest the need for a different framework and practice in which existing epistemic and institutional forms of thinking, political interests and imaginations can explicitly be questioned. This becomes even more obvious when we consider the heterogeneous complexity of post-industrial knowledge in the socio-cultural environment of the Gotthard Base Tunnel exhibition, Jennifer Baichwal's film narrative about the adverse environmental consequences of China's industrial revolution, and Ai Weiwei's social performance *Fairytale* at Documenta 12. In addition, the emerging picture of a global environmental collapse and the threatening rate of consumption far exceeding the planet's load capacity suggest the need for a more anticipatory dynamism based on rigorous reflexive inquiries. To address the cultivation of the technical/instrumental in these inquiries and to implement reflexive forms of learning in different practices whose outcomes impinge on socially complex worlds should therefore be seen as central challenges. These are issues that seem to be part of a larger moral-ecological-political perspective in which cultural work and knowledge are embedded in the framework of implicit collectively shared social (and technical) imaginations as well as diverging views of what is *really* going on.

The task may be first to acknowledge the significance of new learning as well as unlearning dimensions and to redefine them as potentially new forms of (self-)reflexivity. These are, on the one hand, the real challenges of the proposed socio-epistemological-political approach to learning and knowledge which should be applicable to very practical dimensions of cultural work in order to create new spaces and build different realities.

On the other hand, the central challenge is to come to terms with messy practices and the risks of the systems and the contingent frameworks that we have created.[109] The socio-epistemological second-order framework which is based on von Foerster's and Pakman's understanding of human thinking and reflexivity could provide new ways of governing cultural work in order to respond to the growing demand for precaution and, also, to participate in framing and deciding socio-technical issues (cf. Felt, Wynne et al. 2007:54). Furthermore, this framework of ethical, aesthetic, pragmatic and political convergences may support us in remedial efforts to build new (techno-)socio-cultural spaces of knowledge in a democratic knowledge-society.

## Capitalist Cultural Practices

Various parts of this text have problematised the hegemony of capitalist social relations and instrumental orientations. The predominance of industrial society's logic of accumulation and the culture of instrumentalism pose a fundamental challenge to cultural work as they appear to impact the moral foundations of human identity, engagement and sociability. According to Banks, the "moral purpose (…) is driven by the instrumental imperative to make money in a competitive market", and "generating social outcomes or benefits is not usually prioritized" (2007:458). The "gentrification and the commodification of public space" (Ibid.:463) as well as market rationality, the primacy of instrumental values and the "moral vacuum at the heart of cultural industries" (459) seem to under-

---

[109] These frameworks seem to be inadequate to the task of guiding us in a world in which, as Turnbull argues, the "majority of the world's population still live in poverty, the resources that made 'modern civilisation' possible are fast being depleted, and the byproducts of that civilisation threaten to transform the climate of the whole world" (2000:2).

mine the pursuit of a non-instrumental rationale making the problem of commodification and marketisation even worse.[110] Capitalist cultural practices and the "moral vacuum" should therefore be seen as significant challenges not only for professionals such as cultural entrepreneurs, educators, change agents, policy-makers and artists working in different disciplines. At stake is also a different fundamental ethos of learning as well as a different understanding of the spatialisation and dissemination of knowledge that challenge those working as mediators, institutional authorities, decision-makers and arbitrators in the public cultural sector. Instead of taking on more influential and decisive roles, their institutions are increasingly embroiled in markets and run the risk of damaging the long-term reputations they have built up in protecting the public interest.

## Reshaping Existing Knowledge Spaces

The proposed socio-epistemological model of mutually reinforcing reflexive, ethical, pragmatic and political convergences relies on the fragmented public face of the humanities and the arts. It is an open question whether these humanities and arts are at all capable of counteracting the paradigm of instrumental orientations and academic capitalism, which appear to threaten the "real cultivation of the mind" (cf. Blackmore 2001:354). According to Blackmore, the problem lies within the paradigm of higher education itself, which values more its "vocational function in contributing to individual careers or national productivity

---

[110] Although non-instrumental motives may shape practices of cultural entrepreneurship, as Banks argues (2003:466), the potentials to foster "binding effects of sense of place and community obligation" as a focus for "social imperatives that mediate and impose limits around the pursuit of instrumental, profit-seeking goals" (Ibid.), are only inadequately recognised. Banks' insights are based on empirical case studies conducted as interviews with cultural entrepreneurs in Manchester, UK, in the framework, for example, of the "Cultural Industries and the City" project (1980–1999).

rather than intrinsic value" (Ibid.). Both the "power/knowledge crisis" (353) and the "cultivation of the technical" (Nussbaum 2010:23) thus lead to the necessity of putting the distribution of knowledge on the agenda. This is an issue of democratic deliberation in its own right over the "kinds of spaces that we construct in the process of assembling, standardising, transmitting and utilising knowledge" (cf. Turnbull 2000:12) and brings us to the core problem of how to reshape existing knowledge spaces and to create so-called "third" spaces of knowledge.

## "Third" Spaces

A most usefully applicable concept of such "third" spaces is one, I would suggest, that properly acknowledges that knowledge spaces are as much moral as epistemological, political or technical (cf. Turnbull 2000:12). Furthermore, if we agree that "(self-)*reflective* reasoning" (Felt, Wynne et al. 2007:65) is an essential requirement, the emphasis put on the limitations of our knowledge and missed opportunities for learning (Ibid.:26) could become a *real* public issue in such spaces. Moreover, these spaces will support us in exploring the viability of new interactions between audiences and collective matters. We would then be able to reflect upon our authoritative forms of knowledge and the abusive systems we have built, for which we have ourselves to blame. This means that we have the choice to change course and to change the fragmented public face of the humanities and the arts. But this also means that we cannot relax because civil society needs not only to be *informed*, but also to be *mobilised* with regard to sustainable action (cf. Wiegandt 2009:xii).

In view of the failure of (global) capitalism to establish an ecologically sustainable world, it is therefore necessary, I suggest, to begin discussion from within the humanities and arts of how to moderate our behaviours and actions in order to envisage new practices as well as more participatory and "embedded" engagements with an emphasis on the premediation of decisions, purposes and social priorities. Progress along these lines is a matter of sustained democratic dialogue and shared responsibility. As should be clear, there is, however, no direct way out of the actual to the potential with regard to these challenges of social and environmental renewal as well as the present marginalisation of the humanities and arts, which are faced with the relentless growth of science and technology.

Consistent with my emphasis on accommodating the many ambivalent voices of the humanities and arts, I will now focus on the areas resulting from my analysis in which knowledge work and learning in the post-industrial culture may be most important:

- The humanities and arts should be more concerned with our non-knowledge of the world.
- They should be more concerned with the limitations of our knowledge.
- They should become more critical of the knowledges we value.
- They should be more skilled at the spatialisation and dissemination of knowledge.
- They should be more concerned with the consequences of what we do and how we act.
- They should be more concerned with the consequences of science and technology.
- They should be concerned with the dominant idioms of control and disciplinary discourses.
- They should be more concerned with anticipating long-term social impact.

- They should become better at mediating democratic deliberation, more inclusive pluralistic discussions and socio-political priorities.
- They should be concerned with messy practices and instrumental orientations.
- They should be concerned with the unrecognised dimensions of unlearning.
- They should be concerned with new open-ended potentials of collective learning.
- They should be concerned with the social and political action that is contained within cultural work (cf. Banks 2007:460).

This is only a sketch, but it provides a starting point for articulating what the humanities and arts need to do in order to become better at engaging beyond their lateral territories.

A central motivation of this book has been to increase our sensitivity to our collective imagination and alternative collaborative futures. The production of the common and the biopolitical scope of cultural work have emerged as topics in their own right, and I have thus provided inquiries into cultural work around the concept of our "common wealth" in linking some of my arguments to the core ideas in Hardt and Negri's work *Commonwealth* (2009).[111] These issues resonate with Garrett Hardin's[112] paper *The Tragedy of the Commons* (1968)—one of the most-reprinted articles ever to appear in *Science*—which addressed the problem of overpopulation and environmental exploitation as important societal moral and political concerns. My impression is—although moral values may have become an element of broader societal concern (Felt, Wynne et

---

[111] For Hardt and Negri, "the common" stands for "the common wealth of the material world—the air, the water, the fruits of the soil, and all nature's bounty (...)" (2009:viii).
[112] Garrett James Hardin (1915–2003) was a leading American ecologist.

al. 2007:48)—that Hardin's voice (among other salient voices) finds itself expressed today in what Banks refers to as an emerging "critical orthodoxy"—an orthodoxy which concedes that "something is being 'lost' (...) in moral-political or social terms" (2007:458).

## A Practice-Based Epistemology

Given the broad neglect of utopian thinking in the mainstream of critical social science, my aim has been to outline a practice-based epistemology for post-industrial cultural work. But I also wish to reflect upon new educational practices as a catalyst for civic dialogue and cultural change. The conventional notions of cultural work that I have addressed from various angles have been revised and expanded to meet the new requirements of what I propositionally call *cultural knowledge work* (CKW), or briefly *knowledge work* in the post-industrial culture. Foundational contingencies, the problem of uncertainty in scientific knowledge and the self as well as various contradictions have been identified as key challenges to cultivate new ecological and institutional practices. The aim has been to bring these issues from an academic field into a practice-based one.

The idea of cultural work as a union of affective and cognitive capacities is central here to the intrinsic potentials of *cultural knowledge work*. Promoted together with the diverse civic "knowledge-abilities" of different actors, *knowledge work,* which is based on the concept of deeply-ingrained knowledge-practice and knowledge-building intentions (cf. Felt, Wynne et el. 2007:59), could play a central role in cultural production where new forms of knowledge dissemination are becoming increasingly important.

## European Autochthony

The European Commission (EC) 2007 report *Taking European Knowledge Society Seriously,* from which I have quoted extensively, brings me to the final question of how to deal with its key message of global European leadership in science, innovation and governance. The proposal to build a "robust, and sustainable European knowledge-society" and to help Europe lead globally (cf. Felt, Wynne et al. 2007:14, 40, 81) should not only be seen as problematic, reminding us of our Western epistemological and political high stakes, but also what it means to set the "epistemological standard" (Turnbull 2000:6). While what is at stake is public reflection on these issues, the challenge lies in recognising knowledge's "unplanned and messy nature" (Ibid.:1) and coming to terms with what Turnbull calls *"the* millennial problem" (211)—our ways of reasoning and the forms of knowledge we produce as outlined in different parts of my text. This brings me again to the conclusion that properly acknowledging these issues should be seen as the key first step to developing both individual and collective sensitivities as well as to reshaping public discourse around key civic issues.

As social, cultural, economic, political and moral actors, we are all challenged to find new ways of promoting a much wider range of cognitive virtues, to cultivate deeper social values, and to explore new forms of "reality-work" (cf. Law 2009b). I have attempted to address these issues from the perspective of the marginal standing of the humanities and the arts—and cultural work. I have also tried to provide ideas as to how cultural work could become concerned about the whole, which for Beck has become a task (2009:19). To experiment with new forms of civic engagement, to question deeply-entrenched cultural habits, and to explore the disregarded potentials of the humanities and arts thus means that new ways should be found to shape important political values at the heart of social life and creative agency.

## Other Epistemologies

With my theoretical understanding and with what I colloquially call *cultural knowledge work* (CKW), I have attempted to sketch a rationale that could pave the way for other more convenient epistemologies beyond my imaginative shortcomings. Such epistemologies should, I suggest, be oriented towards "socially useful" and "morally progressive" forms of cultural work (Banks 2007:458) as well as beneficial interests to create sustainable life-worlds and new kinds of virtual to-be-built spaces of knowledge. These spaces could then be linked to the unused post-industrial playgrounds of capitalism to resonate with the democratic requirements of knowledge. The proposal would be to address these issues deliberately in search of "blind spots", unknowns, and divergent views (cf. Felt, Wynne et al. 2007:39).

I have attempted to question the theory and narrative of current cultural work practice and to promote the shaping of new practices and new theory. The focus has been on establishing this need and setting out how, through new forms of knowledge dissemination and collective action, we can achieve transformative change. Reflecting on new ways of forging a stronger relationship between the humanities, the arts and civil society should be seen as a pragmatic and urgent challenge.

I do not expect that everything I have discussed in this book will be welcomed and taken up in policy. I feel that our problematic relations with the humanities and arts need to be addressed and that involves a strong degree of challenge and institutional risk. I hope that I have contributed with this book to further thinking and incentives to explore the list of open questions in practice.

The technoscientific globalised world in which we live was not given, but made, and humanly constructed realities are not produced by destiny but may be *remade*. To explicitly combine knowledge practices and cultural work could catalyse development of more effective forms of cultural and scientific learning as well as create new forms of freedom and resistance in a democratic post-industrial culture.

# Bibliography


Adorno, T. (1991). *The Culture Industry: Selected Essays on Mass Culture.* London: Routledge.

Adorno, T. (1981; 1991). Culture industry reconsidered. In: (Bernstein, M., Ed.) *Theodor W. Adorno. The Culture Industry. Selected essays on mass culture.* London, New York: Routledge. pp. 98-106.

Ai Weiwei (2007a). *Fairytales* (Original Project Proposal for Documenta 12, Kassel, Germany). Beijing: Office of Ai Weiwei.

Ai Weiwei (2007b). Interview with Ai Weiwei. *Artkrush*, 63. Available at <http://artkrush.com/128062> downloaded 5 March 2010.

Ai Weiwei, Jocks, H.-N. (2007). Das Märchen von den 1001 Chinesen. Ein Gespräch von Heinz-Norbert Jocks. In: *Kunstforum International,* 187. pp. 428-43.

Ai Weiwei, Birnie Danzker, J.-A. (2008). A Conversation with Ai Weiwei. *Yishu. Journal of Chinese Contemporary Art, 7/4,* July. pp. 10-19.

Ai Weiwei, Jocks, H.-N. (2008). Was heisst Leben? Ein Gespräch von Heinz-Norbert Jocks. In: *Kunstforum International,* 194. pp. 228-39.

Ai Weiwei (2004). A Standpoint without Position. In: *Three Gorges: Displaced Population and Three Gorges: Newly Displaced Population. Liu Xiaodong.* Beijing: China Blue Gallery, China Art Archives & Warehouse. pp. 4-5.

AlpTransit Gotthard Ltd. (Ed.) (2005). *The New Gotthard Rail Link.* Lucerne: AlpTransit Gotthard Ltd.

Altvater, E. (2007). *Das Ende des Kapitalismus wie wir ihn kennen. Eine radikale Kapitalismuskritik.* 5th Edition Münster: Verlag Westfälisches Dampfboot.

Arendt, H. (1958; 1998). *The Human Condition.* Chicago, Illinois: The University of Chicago Press.

Armstrong, J. (2009). The Public Humanist. The Marginalization of the Humanities. *Value Advocate.* Available at <http://www.valleyadvocate.com/blogs/home.cfm?aid=9435> downloaded 1 June 2009.

Ascott, R. (2006). Gardens of Hypotheses and Geomediated Meaning. In: Fiel, W. (Ed.) *Shaun Murray. Disturbing Territories.* Vienna, New York: Springer. pp. 107-13.

Ascott, R. (1990; 2003). Is There Love in the Telematic Embrace? In: Shanken, E. A. (Ed.) *Telematic Embrace. Visionary Theories of Art, Technology, and Consciousness. Roy Ascott.* Berkeley, California: University of California Press. pp. 242-46.

Baker, K. (2007). Manufactured Landscapes. *San Francisco Chronicle,* 20

July. Available at <http://www.sfgate.com/movies/article/FILM-CLIPS-Also-opening-today-2580576.php> downloaded 10 February 2013.

Bakhtin, M. (Holquist, M., Ed.) (1981). *The Dialogical Imagination*. Austin, Texas: University of Texas Press.

Banks, M. (2007). *The Politics of Cultural Work*. Hampshire: Palgrave Macmillan.

Banks, M. (2006). Moral Economy and Cultural Work. *Sociology*, 40/3. pp. 455-72.

Bateson, G. (1979). *Mind and Nature. A Necessary Unity*. Cresskill, New Jersey: Hampton Press Inc.

Beck, U. (2009). *World at Risk*. Cambridge: Polity Press.

Beck, U. (2006). *The Cosmopolitan Vision*. Cambridge: Polity Press.

Beck, U. (1986). *Risikogesellschaft. Auf dem Weg in eine andere Moderne*. Frankfurt a. M.: Suhrkamp Verlag.

Bell, D. (1987). The Post-Industrial Society: A Conceptual Schema. In: Cawkell, A. E. (Ed.) (1986). *Evolution of an Information Society*. London: Aslib.

Bell, D. (1976). *The Cultural Contradictions of Capitalism*. New York: Basic Books.

Bell, D. (1973). *The Coming of Post-Industrial Society. A Venture of Social Forecasting*. New York: Basic Books.

Belliger, A., Krieger, D. (Eds.) (2006). *ANTology. Ein weiterführendes Handbuch zur Akteur-Netzwerk-Theorie*. Bielefeld: Transcript Verlag.

Bender, G. (Ed.) (2001). *Neue Formen der Wissenserzeugung*. Frankfurt a. M.: Campus Verlag.

Berger, B. et al. (2001). *The Humanities in 2010. Alternative Wor(l)ds. Report of the Working Group on the Future of the Humanites to the Board of the Social Sciences and Humanities Research Council of Canada*. Ottawa, Ontario: Social Sciences and Humanities Research Council.

Bińczyk, E. (2008). Looking for Consistency in Avoiding Dualisms. *Constructivist Foundations*, 3/3. pp. 201-7.

Blackmore, J. (2001). Universities in Crisis? Knowledge Economies, Emancipatory Pedagogies, and the Critical Intellectual. *Educational Theory*, 51/3, Summer. pp. 353-70.

Bourdieu, P. (1993). *The Field of Cultural Production: Essays on Art and Literature*. Cambridge: Polity Press.

Bourdieu, P. (1984). *Distinction: A Social Critique of the Judgment of Taste*. London: Routledge and Kegan Paul.

Bourdieu, P., Passeron, J.-C. (1973). Cultural Reproduction and Social Reproduction. In: Brown, R. K. (Ed.) *Knowledge, Education and Cultural Change: Papers in the Sociology of Education*. London: Tavistock.

Capra, F. (2002). *The Hidden Connections: A Science for Sustainable*

*Living*. New York: Anchor Books.

Castells, M. (2000). Information Technology and Global Capitalism. In: Hutton, W., Giddens, A. (Eds.) *Global Capitalism*. New York: The New Press. pp. 52-74.

Choo, Chun Wei, Detlor, B., Turnbull, D. (2000). *Web Work. Information Seeking and Knowledge Work on the World Wide Web*. Dordrecht, Boston, London: Kluwer Academic Publishers.

Clark, K., Holquist, M. (1984). *Mikhail Bakhtin*. Cambridge, Massachusetts: Harvard University Press.

Coggins, D. (2007). Ai Weiwei's Humane Conceptualism. *Art in America,* September. pp. 118-25.

Colonnello, N. (2007). Ai Weiwei—Fairytale. Project for Documenta 12, Kassel, 2007. *Galerie Urs Meile, Beijing–Lucerne*. Available at <http://www.artfacts.net/pdf-files/news/ai%20weiwei%20documenta.pdf> downloaded 21 February 2010.

Collins, D. (1998). Ambiguity and Confusion in the Analysis of the "Knowledge Age". *Journal of Systemic Knowledge Management,* March. Available at <http://www.tlainc.com/article7.htm> downloaded 16 December 2008.

Cotter, H. (2007). Pondering Kassel in some confusion. Documenta 12 makes for quirky viewing. *International Herald Tribune*, 22 June. p. 24.

Despres, C., Hiltrop, J.-M. (1995). Human Resource Management in the Knowledge Age: Current Practice and Perspectives on the Future. *Employee Relations*, 17/1. pp. 9-23.

Drucker, P. F. (1991). The New Productivity Challenge. *Harvard Business Review*, November-December. pp. 69-79.

Drucker, P. F. (1979). Managing the knowledge worker. *Modern Office Procedures*, 24. pp. 12-16.

Drucker, P. F. (1973). *Management: Tasks, Responsibilities and Practices*. New York: Harper and Row.

Dyer, R. (2007). Documenta 12. *Third Text*, 21/6, November. pp. 23-42.

Ellis, A. (2008). The problem with privately funded museums. *The Art Newspaper*, 188. Available at <http://www.theartnewspaper.com/article.asp?id=7509> downloaded 18 June 2008.

Felt, U., Wynne, B. et al. (2007). *Taking European Knowledge Society Seriously*. Report of the Expert Group on Science and Governance to the Science, Economy and Society Directorate, Directorate-General for Research, European Commission. Luxembourg: Office for Official Publications of the European Communities. Available at <http://ec.europa.eu/research/science-society/document_library/pdf_06/european-knowledge-society_en.pdf> downloaded 15 October 2010.

Flusser, V. (1990). *Nachgeschichten. Essays, Vorträge, Glossen*. 1st Edition Düsseldorf: Bollmann Verlag.

Foucault, M. (1978). *Dispositive der Macht. Über Sexualität, Wissen und Wahrheit.* Berlin: Merve Verlag.

Foucault, M. (1972). *The Archaeology of Knowledge & The Discourse on Language.* 1st Edition New York: Pantheon Books.

Fuller, S. (1988; 2002). *Social Epistemology.* 2nd Edition Bloomington, Indianapolis: Indiana University Press.

Gibbons, M. et al. (1994). *The New Production of Knowledge. The Dynamics of Science and Research in Contemporary Societies.* London: Sage.

Giddens, A., Hutton, W. (2000). Fighting Back. In: Hutton, W., Giddens, A. (Eds.) *Global Capitalism.* New York: The New Press. pp. 213-23.

Gerlis, M. (2008). China overtakes France. *The Art Newspaper,* 188, February. p. 39.

Gogan, J. (2002). The Warhol: Museum as Artist: Creative, Dialogic & Civic Practice. Case Study: Without Sanctuary Project. *Americans for the Arts.* Available at <http://www.americansforthearts.org/Animating Democracy/pdf/labs/andy_warhol_museum_case_study.pdf> downloaded 14 July 2009.

Goleman, D. (2009). *Ecological Intelligence.* New York: Broadway Books.

Gorz, A. (1989). *Critique of Economic Reason.* London, New York: Verso.

Gorz, A. (1982). *Farewell to the Working Class: An Essay on Post-Industrial Socialism.* London: Pluto.

Haraway, D. (1997a). *Modest_Witness@Second_Millennium.Female Man©_Meets_OncoMouse™. Feminism and Technoscience.* New York: Routledge.

Haraway, D., Goodeve, T. N. (1997b). How Like a Leaf. A Conversation with Donna Haraway. In: Stocker, G., Schöpf, C. (Eds.) *Fleshfactor. Informationsmaschine Mensch. Ars Electronica Festival 97.* Vienna, New York: Springer. pp. 46-69.

Hardin, G. (1968). The Tragedy of the Commons. *Science,* 162. pp. 1243-248.

Hardt, M., Negri, A. (2009). *Commonwealth.* Cambridge, Massachusetts: The Belknap Press of Harvard University Press.

Hardt, M., Negri, A. (2000). *Empire.* 4th Edition Cambridge, Massachusetts: Harvard University Press.

Harvey, D. (2000). *Spaces of Hope.* Edinburgh: Edinburgh Press.

Horkheimer, M., Adorno, T. W. ( Schmid Noerr, G., Ed.) (1947; 2002). *Dialectic of Enlightenment. Philosophical Fragments.* Stanford, California: Stanford University Press.

Kallas, S., Truttmann, M. (2010). "Das Schweizer Modell ist ein Traum". *Neue Luzerner Zeitung,* 238. p. 3.

Kelloway, E. K., Barling, J. (2000). Knowledge work as organisational behavior. *International Journal of Management Reviews,* 2/3. pp. 287-

304.

Lash, S. (1999). Objects that Judge: Latour's Parliament of Things. Available at <http://translate.eipcp.net/transversal/0107/lash/en> downloaded 14 August 2009.

Lash, S. (1994). Reflexivity and its doubles: structure, aesthetics, community. In: Beck, U., Giddens, A., Lash, S. *Reflexive Modernization: Politics, Tradition and Aesthetics in the Modern Social Order*. Cambridge: Polity Press.

Latour, B. (2005). From Realpolitik to Dingpolitik or How to Make Things Public. In: Latour, B., Weibel, P. (Ed.) *Making Things Public: Atmospheres of Democracy*. Cambridge, Massachusetts; London: MIT Press. pp. 4-31.

Latour, B. (1999a). *Pandora's Hope: Essays on the Reality of Science Studies*. Harvard, Massachusetts: Harvard University Press.

Latour, B. (1999b). The trouble with actor-network theory. *Collaboratory in Critical Security Methods*. Available at <http://www.cours.fse.ulaval.ca/edc-65804/latour-clarifications.pdf> downloaded 12 February. pp. 1-14.

Latour, B. (1993). *We Have Never Been Modern*. Cambridge, Massachusetts: Harvard University Press.

Latour, B., Woolgar, S. (1986). *Laboratory Life. The Construction of Scientific Facts*. Princeton, New Jersey: University Press.

Law, J. (2009a). Actor Network Theory and Material Semiotics. In: Turner, B. S. (Ed.) *The New Blackwell Companion to Social Theory*. Malden, Massachusetts: Blackwell Publishing Ltd. pp. 141-58.

Law, J. (2009b). *Collateral Realities (Version of 29 December 2009)*. Available at <http://heterogeneities.net/publications/Law2009Collateral Realities.pdf> downloaded 8 February 2011. pp. 1-18.

Lefebvre, H. (1974; 2009). *The Production of Space*. Malden, Massachusetts: Blackwell Publishing.

Liu, A. (2004). *The Laws of Cool*. Chicago, Illinois: The University of Chicago Press.

Maasen, S. (1999). *Wissenssoziologie*. 1st Edition Bielefeld: Transcript Verlag.

Massey, D. (2005). *for space*. London: Sage.

Maturana, H., Varela, F. (1987). *The Tree of Knowledge*. Revised Edition Boston: Shambhala Publications.

Mayntz, R., Neidhardt, F., Weingart, P., Wengenroth, U. (Eds.) (2008). *Wissensproduktion und Wissenstransfer. Wissen im Spannungsfeld von Wissenschaft, Politik und Öffentlichkeit*. Bielefeld: Transcript Verlag.

Mäder, M. (2002). Mythos Gotthard. In: Jeker, R. E. (Ed.) *Der längste Tunnel der Welt. Die Zukunft beginnt. Gotthard Basis Tunnel*. Zurich: Werd Verlag. pp.179-97.

McCarthy, D. E. (1996). *Knowledge as Culture. The New Sociology of Knowledge*. London: Routledge.

McDonald, J. (2008). Destruction and creation. *The Sydney Morning Herald*, 17-18 May. pp. 16-17.

McRobbie, A. (2002). Clubs to Companies: Notes on the Decline of Political Culture in Speeded Up Creative Worlds. *Cultural Studies*, 16/4. pp. 516-31.

Merewether, C. (2007). At the Time. In: Meile, U. (Ed.) *Ai Weiwei Works 2004–2007*. Lucerne: Galerie Urs Meile. pp. 173-81.

Mitterer, J. (2001). *Die Flucht aus der Beliebigkeit*. Frankfurt a. M.: Fischer Verlag.

Mitterer, J. (1992). *Das Jenseits der Philosophie*. Vienna: Edition Passagen.

Miyoshi Jager, S. (2007). *On the Uses of Cultural Knowledge*. Carlisle: All Strategic Studies Institute (SSI) Publications.

Münter, U. (2007). Märchen einmal anders. *taz.de*. Available at <http://www.taz.de/?id=archiv&dig=2007/06/11/a0131> downloaded 28 February 2010.

Nagata, A. L. (2004). Promoting Self-Reflexivity in Intercultural Education. *Journal of Intercultural Education*, 8. pp. 139-67.

Nussbaum, M. C. (2010). *Not for Profit. Why Democracy needs the Humanities*. Princeton and Oxford: Princeton University Press.

Opie, B. (2001). Cultural Knowledge and the Conceptions of the Knowledge Society. A paper given at the 13th Round Table of the National Scholarly Communications Forum, Canberra, 9 August 2001. Canberra: Humanities Society of New Zealand.

Pakman, M. (2003). Elements for a Foersterian Poetics in Psychotherapeutic Practice. *Cybernetics & Human Knowing. Heinz von Foerster 1911–2002*, 10/3-4. pp. 107-23.

Pollack, B. (2008a). Making 1,200 Museums Bloom. *Artnews*, 107/3. pp. 123-27.

Pollack, B. (2008b). The Chinese Art Explosion. *Artnews*. Available at <http://www.artnews.com/issues/article.asp?art_id=2542> downloaded 4 March 2010.

Postman, N. (1990). Museum as Dialogue. *Museum News*, 69/5. pp. 55-58.

Puig de la Bellacasa, M. (2011). Matters of care in technoscience: Assembling neglected things. *Social Studies of Science*. 41/1. pp. 85-106.

Report of the Humanities, Science, and Technology Working Group. National Endowment for the Humanities. Available at <http://www.nhalliance.org/neh/policy/hst.pdf> downloaded 22 March 2013.

Ringger, R. (2009). Neuer Kompass für die Welt. *Weltwoche*, 26. p. 20.

Ritzer, G. (2007). *The Coming of Post-Industrial Society*. 2nd Edition New York: McGraw-Hill.

Schmid, A. (2007). Chinesische Kunst auf der Documenta 12. Verstellte Kunst. *Artnet*. Available at <http://www.artnet.de/magazine/features/schmid/schmid08-20-07.asp> downloaded 3 March 2010.

Schnieper, W. (2002). Gotthard Chronik. Eine bewegte Geschichte. In: Jeker, R. E. (Ed.) *Der längste Tunnel der Welt. Die Zukunft beginnt. Gotthard Basis Tunnel.* Zurich: Werd Verlag. pp. 146-51.

Schön, D. A. (1983). *The reflective practioner.* New York: Basic Books.

Searle, A. (2007). 100 days of ineptitude. *The Guardian*, 19 June. pp. 23-26.

Seefranz, C. (2007a). Ai Weiwei. Fairytale. *Documenta 12, Kassel, 16/06–23/09.* Available at <http://www.artfacts.net/pdf-files/inst/ai%20weiwei%20fairytale%20documenta12-en.pdf> downloaded 21 February 2010.

Shanken, E. A. (2003). From Cybernetics to Telematics. The Art, Pedagogy, and Theory of Roy Ascott. In: Shanken, E. A. (Ed.) *Telematic Embrace. Visionary Theories of Art, Technology, and Consciousness. Roy Ascott.* Berkeley, California: University of California Press. pp. 1-94.

Shiva, V. (2000). The World on the Edge. In: Hutton, W., Giddens, A. (Eds.) *Global Capitalism.* New York: The New Press. pp. 112-29.

Sloterdijk, P. (2010). Der Mensch, der Athlet. *Tages-Anzeiger*, 6 February. pp. 35-36.

Smith, D. (2007). Made in China. Jennifer Baichwal and Edward Burtynsky on Their Travels Across Manufactured Landscapes. *Bright Lights Film Journal*, 58. Available at <http://www.brightlightsfilm.com/58/58landscapesiv.html> downloaded 28 February 2009.

Steinmann, N., Favre, P. (2003). *Building a Modern Railway Line in the Gotthard Base Tunnel.* Lucerne: AlpTransit Gotthard AG.

Storl, W.-D. (2007). *Borreliose natürlich heilen.* 2nd Edition Baden and Munich: AT Verlag.

Swidler, A., Arditi, J. (1994). The New Sociology of Knowledge. *Annual Review of Sociology*, 20. pp. 305-29.

Swiss Federal Office of Transport (Ed.) (2007). Through the Alps at 280 km/h. *Alptransit. The Future of Rail*, 2, March. pp. 1-2.

Taylor, S., Bogdan, R. (1998). *Introduction to qualitative research methods: A guidebook and resource.* New York: Wiley.

Tinari, P. (2007). A Kind of True Living. *Artforum International*, Summer. pp. 453-59.

Turnbull, D. (2000). *Masons, Tricksters, and Cartographers. Comparative Studies in the Sociology of Scientific and Indigenous Knowledge.* Amsterdam: Harwood Academic Publishers.

Unterschütz, P. (2002a). *Besucherzentren in Pollegio und Erstfeld.* AlpTransit Gotthard AG.

Unterschütz, P. (2002b). Schutzpatronin der Bergleute. In: Jeker, R. E.

(Ed.) *Der längste Tunnel der Welt. Die Zukunft beginnt. Gotthard Basis Tunnel.* Zurich: Werd Verlag. pp. 157-59.

Vetsch, H.-P. (2002). Lang, länger am längsten. In: Jeker, R. E. (Ed.) *Der längste Tunnel der Welt. Die Zukunft beginnt. Gotthard Basis Tunnel.* Zurich: Werd Verlag. pp.152-56.

Von Foerster, H., Pörksen, B. (1998). *Wahrheit ist die Erfindung eines Lügners. Gespräche für Skeptiker.* Heidelberg: Carl-Auer-Systeme Verlag.

Von Foerster, H. (1984). *Observing systems.* Seaside, California: Intersystems.

Von Foerster, H. (1973). On constructing reality. In: Preiser, F. E. (Ed.) *Environmental design research,* 2. pp. 35-46.

Walsham, G. (1997). Actor-Network Theory and IS Research: Current Status and Future Prospects. In: Allen, S. L., Liebenau, J., DeGross, J. I. (Eds.) *Information Systems and Qualitative Research.* London: Chapman & Hall. pp. 466-79.

Welzer, H. (2008). *Klimakriege. Wofür im 21. Jahrhundert getötet wird.* Frankfurt a. M.: Fischer Verlag.

Wertheim, M. (1997). *Phythagoras' Trousers. God, Physics, and the Gender Wars.* New York, London: W. W. Norton & Company.

Willis, P. (2005). Invisible Aesthetics and the Social Work of Commodity Culture. In: Inglis, D., Hughson, J. (Eds.) *The Sociology of Art.* Basingstoke: Palgrave Macmillan. pp. 73-86.

Willke, H. (1998). Organisierte Wissensarbeit. *Zeitschrift für Soziologie,* 27. pp. 161-77.

Yang Jiechang, Köppel-Yang, M. (2007). René Block's Waterloo: Some Impressions of documenta 12, Kassel. *Yishu. Journal of Chinese Contemporary Art,* 6/4, December. pp. 93-100.

# List of Illustrations and Sources

**Fig. 1.1**

Cartoon *Economy of Knowledge*. Courtesy: Gabi Kopp and author; 2008.

**Fig. 2.1**

*Cultural Work and "Cultural Knowledge Work"*. Courtesy: author; 2013.

**Fig. 2.2**

*InfoCenter Erstfeld,* Gotthard Base Tunnel construction site.
Photo: author; 2008.

**Fig. 2.3**

*The European High-Speed Rail Network in 2020*. The New Gotthard
Rail Link, 2005, p. 2. Courtesy: AlpTransit Gotthard Ltd. (Ed.).

**Fig. 2.4**

*Saint Barbara*, InfoCenter Erstfeld, Gotthard Base Tunnel construction
site. Photo: author; 2008.

**Fig. 2.5**

*William Tell and exhibition visitors,* InfoCenter Erstfeld, Gotthard Base
Tunnel construction site. Courtesy: Uri Tourismus AG and AlpTransit
Gotthard Ltd.; 2013.

**Fig. 2.6**

*Video Installation*, InfoCenter Erstfeld, Gotthard Base Tunnel construction site. Photo: author; 2008.

**Fig. 2.7, 2.8, 2.9**

*InfoCenter Erstfeld*, Gotthard Base Tunnel construction site.
Photos: author; 2008.

**Fig. 2.10**

<http://www.alptransit.ch/en/visit-us/visit-us.html>
downloaded 20 January 2013. Courtesy: AlpTransit Gotthard Ltd.

**Fig. 2.11**

*Post-Industrial Cultural Work, (Self-)Reflexivity, Imagination, Learning.*
Courtesy: author; 2013.

**Fig. 2.12–2.15; 2.18–2.24; 2.26; 2.27–2.30; 2.35–2.40**

Film Stills from *Manufactured Landscapes* (2006);
Courtesy: Jennifer Baichwal.

**Fig. 2.12, 2.13, 2.15, 2.26**

Film Stills from *Manufactured Landscapes* (2006). Black and white footage by Jeff Powis.
© Edward Burtynsky <http://www.edwardburtynsky.com>.

**Fig. 2.16, 2.17, 2.25, 2.31, 2.33, 2.34**

Photos © Edward Burtynsky <http://www.edwardburtynsky.com>;
Courtesy: Nicholas Metivier Gallery, Toronto.

**Fig. 2.41**

*Post-Industrial Cultural Knowledge Work.* Courtesy: author; 2013.

**Fig. 2.42**

Cover page of the booklet and concept by Ai Weiwei for *"Fairytale" 2007.* Courtesy: Ai Weiwei; Leister Foundation, Switzerland; Erlenmeyer Stiftung, Switzerland; Galerie Urs Meile, Beijing–Lucerne.

**Fig. 2.43**

*The five flights needed to transport the 1,001 Chinese participants at Documenta 12, 2007, from China to Germany and back to China.* Courtesy: author; 2010.

**Fig. 2.44**

*"Fairytale" 2007,* 1001 Qing Dynasty (1644–911) wooden chairs, exhibition view.
<http://www.flickr.com/photos/70839462@N00/2804254754>
downloaded 7 March 2010; Courtesy: Adrian Koss; Ai Weiwei;
Galerie Urs Meile, Beijing–Lucerne.

**Fig. 2.45, 2.46**

*"Template" 2007,* wooden doors and windows from destroyed Ming and Qing Dynasty houses (1368–1911); <http://www.chinesische-gegenwartskunst.de/pages/ausstellungen/ai-weiwei-documenta-12.php>
downloaded 7 March 2010.
Courtesy: Nadine Dinter – Photography (Fig. 2.44); Ai Weiwei;
Galerie Urs Meile, Beijing–Lucerne (Fig 2.44, 2.45).

**Fig. 2.47**

*The Cultural and Economic Actors behind "Fairytale".*
Courtesy: author; 2013.

**Fig. 2.48**

*48 Seiten Kunst Edition.* Vanity Fair, 24, 2007, p. 18. Courtesy: Sven Paustian and Vanity Fair. Ai Weiwei posing in a saucepan for the media in the "Fairytale" kitchen at Documenta 12, Kassel, Germany.

**Fig. 2.49**

*48 Seiten Kunst Edition*. Vanity Fair, 24, 2007, p. 22. Courtesy: Sven Paustian and Vanity Fair. Ai Weiwei posing with a cigarette butt in his navel for the media at Documenta 12, Kassel, Germany.

**Fig. 2.50**

*"Fairytale People"* 2007, c-prints, each 100 x 100 cm. Courtesy: Ai Weiwei; Leister Foundation, Switzerland; Erlenmeyer Stiftung, Switzerland; Galerie Urs Meile, Beijing–Lucerne.

**Fig. 2.51**

<http://www.artnet.de/magazine/features/schmid/schmid08-20-07.asp> downloaded 7 March 2010. Courtesy: Andreas Schmid.

**Fig. 2.52, 2.53**

"Fairytale" participants prior to their departure from Kassel, Germany. Courtesy: Julia Zimmermann; Ai Weiwei; Leister Foundation, Switzerland; Erlenmeyer Stiftung, Switzerland; Galerie Urs Meile, Beijing–Lucerne.

**Fig. 2.54**

"Fairytale" participants waiting prior to their departure from Kassel, Germany. Courtesy: Ai Weiwei; Leister Foundation, Switzerland; Erlenmeyer Stiftung, Switzerland; Galerie Urs Meile, Beijing–Lucerne; 2007.

**Fig. 2.55**

"Fairytale" participants (fourth group) prior to their departure from Kassel, Germany. Courtesy: Ai Weiwei; Leister Foundation, Switzerland; Erlenmeyer Stiftung, Switzerland; Galerie Urs Meile, Beijing–Lucerne; 2007.

**Fig. 2.56**

Cartoon *Kunstmarkt extra*. Frankfurter Allgemeine Zeitung, 85, 12 April 2007, p. K4. Courtesy: Volker Reiche.

**Fig. 2.57**

*Shanghai Uni statt Tokio Hotel*. Hessische/Niedersächsische Allgemeine, 23 June 2007. Courtesy: Wilhelm Ditzel and HNA.

**Fig. 2.58**

*Zum Abschluss Fussball. Ai Weiwei's Chinesen kickten gegen Nordstadtmannschaft*. Hessische/Niedersächsische Allgemeine, 23 June 2007; Courtesy: Wilhelm Ditzel and HNA.

**Fig. 2.59**

*"Cooking for China": Kreatives Kochen. Teil des Wei Wei-Projektes.* Fränkische Landeszeitung, 150, 3 July 2007. Courtesy: Radek Seidl.

**Fig. 4.1**

*Knowledge Work in the Post-Industrial Culture.* Courtesy: author; 2013.

**Fig. 4.2**

*Knowledge Work in the Post-Industrial Culture (Ethics, Aesthetics, Pragmatics, Politics).* Courtesy: author; 2013.

**Fig. 4.3**

*Knowledge Work in the Post-Industrial Culture (Language, Reflexivity, Temporality).* Courtesy: author; 2013.

# Index

# About the Author

René Stettler is the founder of the Neue Galerie Luzern (1987), the Swiss Biennial on Science, Technics + Aesthetics (1994) and the NGL – SAA Neue Galerie Luzern – Swiss Academic Association (2013). The Swiss Biennial on Science, Technics + Aesthetics has been a forum for discussion of major topics such as Brain–Mind–Culture (1995), Liquid Visions (1997), Frontier Communication: Human Beings, Apes, Whales, Electronic Networks (1999), The Enigma of Consciousness (2001), Consciousness and Teleportation (2005), Consciousness and Quantum Computers (2007), and The Large, the Small and the Human Mind (Part 1 in 2010, and Part 2 in 2012) by internationally acclaimed speakers such as the British mathematician Sir Roger Penrose, the Austrian quantum physicist Anton Zeilinger, the German chaos theorist Otto E. Rössler, the French sociologist Bruno Latour, the Austrian-American ecologist Fritjof Capra and the Austrian philosopher Josef Mitterer.

Stettler received his Ph.D. from the University of Plymouth for the thesis *The Politics of Post-Industrial Cultural Knowledge Work* (2011) under the guidance of Roy Ascott and David Turnbull. His research interests include the theory of cultural work, cultural and scientific learning, the spatialisation and dissemination of knowledge, the construction of public spaces of knowledge for civic reflection, cultural policy, and knowledge politics. His areas of interest are the sociology of knowledge, the socio-epistemological-political responsibility of cultural work, and the social and educational challenges in the face of industrial society's logic of accumulation, market rationality and instrumentalism. In the past Stettler has supervised many bachelor and master theses at the University of Applied Sciences and Arts in Lucerne, Switzerland. Since 2013 he has been the director of studies of a Swiss-based Ph.D. research programme in collaboration with the University of Plymouth and the Planetary Collegium, Plymouth, UK.